CONTROL AND DYNAMIC SYSTEMS

Advances in Theory and Applications

Volume 47

CONTRIBUTORS TO THIS VOLUME

WARREN J. BOE
P. BUTRYN
LUIS C. CATTANI
CHUNHUNG CHENG
ALAN A. DESROCHERS
FRANK DICESARE
PAUL J. EAGLE
JOONG-IN KIM
VASSILIS S. KOUIKOGLOU
J. KRAMER
ANDREW KUSIAK
J. MAGEE
Y. NARAHARI
YANNIS A. PHILLIS
R. RAM
IN-KYU RO
M. SLOMAN
EDWARD SZCZERBICKI
N. VISWANADHAM
J. M. WILSON
BIN WU
S. DAVID WU
RICHARD A. WYSK

CONTROL AND DYNAMIC SYSTEMS

ADVANCES IN THEORY AND APPLICATIONS

Edited by

C. T. LEONDES

School of Engineering and Applied Science
University of California, Los Angeles
Los Angeles, California
and
College of Engineering
University of Washington
Seattle, Washington

VOLUME 47: MANUFACTURING AND
AUTOMATION SYSTEMS:
TECHNIQUES AND TECHNOLOGIES
Part 3 of 5

ACADEMIC PRESS, INC.
Harcourt Brace Jovanovich, Publishers
San Diego   New York   Boston
London   Sydney   Tokyo   Toronto

ACADEMIC PRESS RAPID MANUSCRIPT REPRODUCTION

Academic Press, Inc.
San Diego, California 92101

United Kingdom Edition published by
Academic Press Limited
24–28 Oval Road, London NW1 7DX

Library of Congress Catalog Number: 64-8027

International Standard Book Number: 0-12-012747-4

PRINTED IN THE UNITED STATES OF AMERICA
91 92 93 94   9 8 7 6 5 4 3 2 1

CONTENTS

CONTRIBUTORS

Numbers in parentheses indicate the pages on which the authors' contributions begin.

Warren J. Boe (399), *Intelligent Systems Laboratory, Department of Industrial Engineering, The University of Iowa, Iowa City, Iowa 52242*

P. Butryn (325), *BP Research, BP International Ltd., Sunbury Research Centre, Sunbury on Thames, Middlesex TW16 7LN, England*

Luis C. Cattani (357), *Department of Mechanical Engineering, University of Detroit Mercy, Detroit, Michigan 48221*

Chunhung Cheng (399), *Intelligent Systems Laboratory, Department of Industrial Engineering, The University of Iowa, Iowa City, Iowa 52242*

Alan A. Desrochers (121), *Electrical, Computer, and Systems Engineering Department, Center for Intelligent Robotic Systems for Space Exploration, Rensselaer Polytechnic Institute, Troy, New York 12180*

Frank Dicesare (121), *Electrical, Computer, and Systems Engineering Department, Center for Manufacturing Productivity and Technology Transfer, Rensselaer Polytechnic Institute, Troy, New York 12180*

Paul J. Eagle (357), *Department of Mechanical Engineering, University of Detroit Mercy, Detroit, Michigan 48221*

Joong-In Kim (241), *CIMS Laboratory, Department of Industrial Engineering, Hanyang University, Seoul 133-791, Korea*

Vassilis S. Kouikoglou (1), *Department of Production Engineering and Management, Technical University of Crete, Chania 731 00, Greece*

J. Kramer (325), *Department of Computing, Imperial College of Science Technology and Medicine, London SW7 2BZ, England*

Andrew Kusiak (173, 221, 399), *Intelligent Systems Laboratory, Department of Industrial Engineering, The University of Iowa, Iowa City, Iowa 52242*

J. Magee (325), *Department of Computing, Imperial College of Science Technology and Medicine, London SW7 2BZ, England*

Y. Narahari (77), *Department of Computer Science and Automation, Indian Institute of Science, Bangalore 560 012, India*

Yannis A. Phillis (1), *Department of Production Engineering and Management, Technical University of Crete, Chania 731 00, Greece*

R. Ram (77), *Department of Computer Science and Automation, Indian Institute of Science, Bangalore 560 012, India*

In-Kyu Ro (241), *CIMS Laboratory, Department of Industrial Engineering, Hanyang University, Seoul 133-791, Korea*

M. Sloman (325), *Department of Computing, Imperial College of Science Technology and Medicine, London SW7 2BZ, England*

Edward Szczerbicki (173), *Intelligent Systems Laboratory, Department of Industrial Engineering, The University of Iowa, Iowa City, Iowa 52242*

N. Viswanadham (77), *Department of Computer Science and Automation, Indian Institute of Science, Bangalore 560 012, India*

J. M. Wilson (383), *Loughborough University of Technology, Loughborough, Leicestershire LE11 3TU, England*

Bin Wu (281), *Department of Manufacturing and Engineering Systems, Brunel University, Uxbridge, Middlesex UB8 3PH, England*

S. David Wu (51), *Department of Industrial Engineering, Lehigh University, Bethlehem, Pennsylvania 18015*

Richard A. Wysk (51), *Department of Industrial Engineering, Texas A&M University, College Station, Texas 77843*

PREFACE

At the start of this century, national economies on the international scene were, to a large extent, agriculturally based. This was, perhaps, the dominant reason for the protraction, on the international scene, of the Great Depression, which began with the Wall Street stock market crash of October 1929. In any event, after World War II the trend away from agriculturally based economies and toward industrially based economies continued and strengthened. Indeed, today, in the United States, approximately only 1% of the population is involved in the agriculture industry. Yet, this small segment largely provides for the agriculture requirements of the United States and, in fact, provides significant agriculture exports. This, of course, is made possible by the greatly improved techniques and technologies utilized in the agriculture industry.

The trend toward industrially based economies after World War II was, in turn, followed by a trend toward service-based economies; and, in fact, in the United States today roughly 70% of the employment is involved with service industries, and this percentage continues to increase. Nevertheless, of course, manufacturing retains its historic importance in the economy of the United States and in other economies, and in the United States the manufacturing industries account for the lion's share of exports and imports. Just as in the case of the agriculture industries, more is continually expected from a constantly shrinking percentage of the population. Also, just as in the case of the agriculture industries, this can only be possible through the utilization of constantly improving techniques and technologies in the manufacturing industries. As a result, this is a particularly appropriate time to treat the issue of manufacturing and automation systems in this international series. Thus, this is Part 3 of a five-part set of volumes devoted to the most timely theme of "Manufacturing and Automation Systems: Techniques and Technologies."

The first contribution to this volume is "Techniques in Modeling and Control Policies for Production Networks," by Yannis A. Phillis and Vassilis S. Kouikoglou. Large, interconnected systems are widely used in manufacturing. A production system consists of a set of workstations and a transfer mechanism to handle the part flows. Workstations are parallel configurations of machines that perform specific types of operations. Due to deterioration, machines are prone to failures that occur at random instants. A failed machine remains inoperable for some period of random duration until it is repaired and ready to operate again. Various types of raw parts are successively loaded to machines and are finally transformed into finished products. Due to machine breakdowns and different processing rates, the part flow between adjacent workstations may be reduced or interrupted. To avoid this, buffers of finite capacity are placed in front of each workstation to provide temporary storage space for workparts. Optimal design and control of production systems are problems of major economic importance in modern industries and depend on the development of efficient analytic tools to evaluate the performance of candidate optimization strategies. Analysis is required in designing a new system, in expanding an existing one, and in short-term production control, with the objective of achieving a desired performance while keeping investment and operational costs low. Performance measures, such as average throughput rates and product cycle times, are estimated by exploiting the model of the system. This contribution provides an in-depth treatment of techniques for modeling and the development of optimal control policies for production networks. Because of the great importance of this issue in manufacturing sytems, this is a particularly appropriate contribution with which to begin this volume.

The next contribution is "Scheduling, Optimization, and Control in Automated Systems," by S. David Wu and Richard A. Wysk. The performance of an automated manufacturing system not only relies on state-of-the-art machinery but also on the effective planning and control of day-to-day operations. Of critical importance is the scheduling of production operations and the control of planned activities. Traditional methods to shop scheduling typically involve the generation of a static, precomputed schedule using optimization or heuristic methods. The static schedule is generated based on certain managerial or operational objectives subject to shop constraints. In real-world applications, however, the moment a schedule is released to the shop, various uncertainty aspects of the system will render this schedule nothing but a far-reaching guideline for shop operations. As a result, the planning of activities in the shop relies heavily on human intervention. This contribution is an in-depth treatment of the broadly complex and interacting issues of scheduling, optimization, and control in production systems.

The next contribution is "Performability of Automated Manufacturing Systems," by N. Viswanadham, Y. Narahari, and R. Ram. Automated manufacturing systems (AMSs) can be regarded as flexible, degradable, fault-tolerant systems. To evaluate such systems, combined measures of performance and reliability, called performability measures, are needed. The most important performability measures in the AMS context are related to throughput and manufacturing lead time (MLT), since high productivity and low lead times are prime features determining the competitiveness of AMSs. Performability has a strong relation to the notion of flexibility in manufacturing systems. Performability modeling is an active research topic in the area of fault-tolerant computing systems. Performability studies are of great interest in the AMS context also, since performability enables one to quantify flexibility and competitiveness of a manufacturing system. Steady-state performability, interval performability, and distribution of performability with respect to throughput and MLT measures are considered. Through several illustrative examples and numerical results, the importance of performability modeling and evaluation in AMS design is brought out.

The next contribution is "Modeling, Control, and Performance Analysis of Automated Manufacturing Systems Using Petri Nets," by Frank Dicesare and Alan A. Desrochers. The utilization of Petri nets has become more important in recent years because it can solve problems that cannot be modeled using queueing theory, while avoiding the time-consuming, trial-and-error approach of simulation. The modeling problem is characterized by concurrent and asynchronous events that are typical for such discrete event dynamic systems. Petri nets are well suited for modeling manufacturing systems because they capture the precedence relations and interactions among these events. In addition, a strong mathematical foundation exists for describing these nets. This allows a qualitative analysis of such system properties as deadlock, conflict, and boundedness. The Petri net model can also be used as the basis of a real-time controller for a manufacturing system. The flow of tokens through the net establishes the sequence of events to carry out a specific task, such as the manufacturing of a particular part type. Petri net controllers have been used in factories in Japan and Europe. Because of the power of the utilization of Petri net techniques in manufacturing systems, this contribution is an essential element of this set of volumes.

The next contribution is "*Evalcon*: A System for Evaluation of Concurrency in Engineering Design," by Andrew Kusiak and Edward Szczerbicki. Design undergoes evaluation at all of its stages. Evaluation of design from the perspective of various life-cycle attributes, such as manufacturability, assemblability, reliability, and so on, is named a concurrency evaluation. The concurrency evaluation process in routine design should assist a designer in specification, selection, and synthesis of parts. The main difficulty associated with the evaluation process is caused by the incompleteness and uncertainty of information. To facilitate the routine design and to allow for incremental concurrency evaluation as the design evolves, interactive design support techniques are presented in this contribution, and their essential importance in the product design process in manufacturing systems is illustrated by examples that are presented in this contribution.

The next contribution is "Concurrent Engineering: Design of Assemblies for Schedulability," by Andrew Kusiak. The traditional design of assemblies (products) has relied on an iterative approach. The main difficulty with the iterative approach is that it is time consuming and many iterations are required before a design project is completed. In order to reduce the design cycle, concurrent design has been introduced. The basic idea of concurrent design is to shorten the time horizon in which the design constraints, such as schedulability, manufacturability, quality, reliability, maintainability, and so on, are introduced. This contribution focuses on the schedulability constraint. It is shown that assembly design has a significant impact on the complexity of the scheduling problem and its solution quality.

The next contribution is "Multi-Criteria Optimization and Dynamic Control Methods in Flexible Manufacturing Systems," by In-Kyu Ro and Joong-In Kim. There are many methods used to solve scheduling and operational control problems in flexible manufacturing systems (FMSs). These methods are mathematical programming techniques (optimization and heuristic methods), queueing networks, computer simulation, Petri nets, and expert systems, etc. Most FMS models are based on one out of the several methods listed above and they are single-objective oriented techniques. However, the combination of several methods may give better results than a single one and decision makers usually consider multiple objectives. On the other hand, most FMS models consider the scheduling for jobs and automated guided vehicles (AGVs) separately. However, jobs and AGVs are closely related, and thus their simultaneous optimal scheduling is essential in manufacturing systems. This contribution presents techniques for these important issues and illustrates them with numerous examples.

The next contribution is "Computer-Aided Manufacturing Systems Design: Framework and Tools," by Bin Wu. In this contribution Wu presents a framework of computer-aided manufacturing systems design (CAMSD) and discusses the tools and methodologies that can be used in an integrated environment. Two distinct approaches can be taken to system design, i.e., the top-down and the bottom-up. The former starts with a set of objectives and then creates a system model that fits the intended purpose. This approach gives little consideration to the current system being operated. Although preferable from the designer's point of view, it can result in a design that requires a total replacement of the current system and, therefore, heavy capital investment. The other approach bases its consideration mostly on the existing system, producing designs that require less capital investment. However, since it is unlikely that any manufacturing unit was initially designed to cope with later structural and functional alterations, the options available could be constrained and ideas severely limited. The approach presented in this contribution is a hybrid of these two approaches. Dr. Wu is one of the leading research workers in manufacturing systems on the international scene, and he has written one of the most important texts on this subject (*Fundamentals of Manufacturing Systems Design and Analysis*, Chapman and Hall, London, 1991). Therefore, his contribution on the greatly important subject of

CAMSD constitutes a most welcome element of this series of volumes.

The next contribution is "Software Configuration Techniques in Operational Systems," by M. Sloman, J. Kramer, J. Magee, and P. Butryn. There has been considerable research into programming languages that emphasize modularity and the reuse of components (e.g., ADA, SR, Argus, NIL, and Emerald). However, there has been comparatively little work on languages that emphasize the separation of programming individual components (programming in the small) from the specification and construction of distributed systems from predefined, reusable components (programming in the large). The various languages mentioned above do have some support for combining separately compiled modules, but they treat the resultant system as a single large program. The specification of a current system is buried within the current state of a program, which makes it difficult to modify or extend the system. The diversity of the software components useful in large distributed systems, such as manufacturing systems, implies that appropriate, state-of-the-art programming languages should be used. For instance, high-level procedural languages (e.g., C, Pascal, and Modula2) should be used for the real-time control and monitoring, object-oriented languages (e.g., Smalltalk and C++) for man–machine interfaces, artificial intelligence languages (e.g., Prolog and Lisp) for the expert systems and knowledge bases needed for the advanced intelligent (i.e., decision-making) components, and Fortran for the existing numerical analysis packages. Although the rewards are great, the use of heterogeneous programming languages exacerbates the problems of integration, as they may make use of incompatible communication mechanisms and data representations. This contribution presents a configuration approach to integrating heterogeneous software components that communicate and interact in a distributed computer automation system.

The next contribution is "Software for Dynamic Tool Modeling in Manufacturing and Automation," by Luis C. Cattani and Paul J. Eagle. The number of robot applications has increased dramatically since 1980 and the predominant application has been spot welding in the automotive industry. Unfortunately, spot welding tools or guns typically impart large loading on robots and have made robot reliability an important issue. Mechanical failures have been identified with the symptoms of excessive backlash and harmonic drive failures. Excessive backlash can cause poor weld quality while the harmonic drive failures can cause downtime. In spot welding applications, robot overloading is caused by the weld guns producing large reaction torques on the robot, exceeding the robot design capacity. An often-overlooked source of robot loading occurs due to dynamic effects. This is due to the difficulty of calculating the moments of inertia of the welding guns. This complex geometry is a result of the sizes and shapes of the welding tools needed to operate on car bodies. To address this problem, a simple solid modeling software called TANGO-UD has been developed by Cattani and Eagle, and it is presented in this contribution. It calculates and analyzes solid properties (mass, volume, moments of inertia, products of inertia, and radius of gyration) of welding guns and other mechanical components. The software has a geometric and material definition language, a parser and compiler, a mass properties calculator, a mass properties analyzer, and the ability to display the wire frame model of the object being analyzed using the AutoCAD graphics software.

The next contribution is "Techniques for Optimal Operation Allocation Methods in Manufacturing Systems—A Review of Non-Probabilistic Approaches," by J. M. Wilson. The field of optimal operation allocation methods has been a rich source for theoretical research and applications development in the manufacturing systems subject area. Starting from the early work of A. S. Manne in the 1950s, the subject of optimal allocation has moved forward with great rapidity in the 1960s and 1970s. Developments after the 1970s might have seemed to be entering into a rather arid phase, with more research than applications to use it. However, two major changes took place that put new momentum into research and applications, and this momentum has been maintained into the 1990s and is continuing. The two developments were (1) the classification of the complexity of algorithms for many types of problem and (2) the introduction of flexible manufacturing systems. This contribution provides an in-depth treatment of optimal operation allocation techniques.

The final contribution to this volume is "A Branch-and-Bound Algorithm for Solving the Machine Allocation Problem," by Chunhung Cheng, Andrew Kusiak, and Warren J. Boe. Group technology (GT) is concerned with the formation of part families and machine cells. The result of grouping machines and parts may lead to either a physical or logical machine layout. The physical machine layout implies rearrangement of machines on the shop floor. The logical machine layout requires virtual grouping of machines and therefore does not alter the position of machines on a shop floor. In this contribution an in-depth treatment of the problem of allocating machines to cells (machine layout) in an optimum manner is presented, and, because of the importance of this issue in manufacturing systems, it is an appropriate contribution with which to conclude this volume.

This volume is a particularly appropriate one as the third of a companion set of five volumes on techniques and technologies in manufacturing and automation systems. The authors are all to be congratulated for their superb contributions, which will provide a uniquely significant reference source for workers on the international scene for years to come.

TECHNIQUES IN MODELING
AND CONTROL POLICIES
FOR PRODUCTION NETWORKS

YANNIS A. PHILLIS
VASSILIS S. KOUIKOGLOU

Department of Production Engineering & Management
Technical University of Crete
Chania 731 00, Greece

I. INTRODUCTION

Large, interconnected production systems are widely used in manufacturing. A production system consists of a set of workstations and a transfer mechanism to handle the part flows. Workstations are parallel configurations of machines which perform specific types of operations. Due to deterioration, machines are prone to failures which occur at random instants. A failed machine remains inoperable for some period of random duration until it is repaired and ready to operate again.

Various types of raw parts are successively loaded to machines and are finally transformed into finished products. Due to machine breakdowns and different processing rates, the part flow between adjacent workstations may be reduced or interrupted. To avoid this, buffers of finite capacity are placed in front of each workstation providing temporary storage space for workparts.

Optimal design and control of production systems are problems of major economic importance in modern industries and depend on the development of efficient analytic tools to evaluate the performance of candidate optimization strategies. Analysis is required in designing a new system, expanding an existing one, and in short-term production control, with the objective of achieving a desired performance while keeping investment and operational costs low. Performance measures, such as average throughput rates and product cycle times, are estimated by exploiting the model of the system.

Once a computationally efficient and accurate estimation method is developed, it can be used to evaluate alternative designs and control policies.

CONTROL AND DYNAMIC SYSTEMS, VOL. 47

Typical decision parameters are the following:
a)   Number of parallel machines in workstations;
b)   Production rates of machines;
c)   Number of repairmen in the system;
d)   Repair allocation policies for systems with limited number of repairmen;
e)   Buffer capacities.

The problem then becomes one of determining the decision parameters to maximize the profit subject to capital investment and physical limitations.

The problems of analysis and optimization have been studied by many researchers. Yet, they remain in general unsolved, due to the inherent complexity of production systems and the lack of an exploitable model for their description.

Markovian models for queueing systems have been used to analyze the steady-state characteristics of production systems. These models were developed to study various stochastic processes arising in computer, communication, and traffic networks. At any instant of the production period the system state is determined by the number of parts in buffers and the operational conditions of machines. Closed form solutions for the steady-state probabilities exist for networks with infinite storage capacities and reliable machines having exponential processing times [1-4]. For production lines with unreliable machines and zero or infinite storage spaces it is possible to calculate the throughput rate in terms of the stand-alone throughput rates of machines [5,6].

Systems with limited storage are difficult to study due to the enormity of the state vector. However, two-stage systems with an intermediate buffer have been efficiently analyzed in [7-12]. The basic assumptions adopted therein include phase-type processing times or unreliable machines with constant or exponential processing times, continuous or discrete-part flow, and possible scrapping of parts. By exploiting the difference equations of the corresponding Markov model the steady-state probabilities are found to have a sum-of-products form.

Larger systems can be approximated using decomposition and aggregation techniques [13-20]. The idea is analogous to the Norton equivalent in electrical networks. The system is divided into two subsystems, one being upstream and the other downstream a buffer. Each subsystem is represented by an equivalent machine. Hence the whole system is approximated by a simple two-stage one. The operating characteristics of each equivalent machine are determined by carrying out further decompositions of the corresponding subsystem.

Another approximate method is the mean value analysis which was developed for the study of computer networks [21,22]. The solution procedure is based on the result that various performance measures of a queueing system can be expressed in terms of average buffer levels and waiting times in queues. Little's formula is an analogous result for two-stage systems. More general production networks can be efficiently approximated [23].

A minimax algebra model proposed recently in [24] has been applied to deterministic systems. The method takes into account the timing constraints which rule the order of part arrivals and departures from machines. The main advantage of this approach is that it can be applied to both transient and steady states. Yet only a few results have been reported concerning unreliable production networks [25].

While these methods require small computation times, they are based on rather restrictive assumptions such as exponential-type distributions for processing, failure, and repair times of machines and steady-state operation. In addition, most of them suffer from the curse of dimensionality. In fact, the state vector consists of the probabilities of all possible states which may be visited by the system and its dimension grows exponentially with system size. For a completely general analysis of production systems, simulation is the inevitable alternative.

A simulation model describes a possible evolution of the production process. Such models known also as piece-by-piece or brute-force, use a computer program to observe the workparts as they move sequentially through the production system. The system state consists partly of discrete states, such as buffer levels and machine operating conditions, and continuous states, such as machine residual times-to-failure, repair, or production of a workpart. When a continuous-state variable reaches zero, one or more discrete states are modified. In a piece-by-piece simulation model the discrete-state transitions, arrivals or departures from machines are marked as elementary events. To each machine and buffer a candidate event is assigned along with the corresponding time of occurrence. The smallest event time determines the next state transition. After executing a next event, the state vector is modified according to a set of rules imposed by the functional characteristics of the production system. Using a random number generator together with appropriate distribution functions, the times of occurrence of new events are computed. The process then repeats.

This technique is called discrete-event simulation because it views the production system as a discrete event dynamic system (DEDS). In the last two decades a number of general purpose simulation languages have been developed such as SLAM, SIMSCRIPT and GPSS, which are widely used in modeling DEDS's.

Various researchers have also used simulation in optimization. Early studies that appeared in the literature examine the problem of buffer capacity allocation and control of production lines [26-29]. However, a common disadvantage of simulation versus analytical models is their long run times. There are two reasons behind this, the time consuming piece-by-piece tracking of the system function and the need for several independent runs for obtaining the desired accuracies for various candidate designs. Efforts towards combining analytic and simulation models have achieved considerable reduction in computation times. In [30] a unifying definition and classification of such hybrid models is given.

Perturbation analysis (PA) is a hybrid technique that calculates the partial derivatives of performance measures of a DEDS with respect to decision parameters. By using PA during a simulation run one can predict the evolution of a perturbed system resulting from a small change in a parameter without needing additional simulation. This is accomplished by applying a number of simple perturbation rules. The corresponding computational savings versus brute-force experimentation are proportional to the number of decision parameters. The method has been applied together with gradient optimization procedures in allocation of buffer capacities for production lines and machine average production rates for queueing networks [31-34]. Extensions to PA proposed in [35,36] yield accurate sensitivity estimates with respect to discrete-decision parameters in a number of experiments.

Recently, a new hybrid event-driven method has been developed which increases the execution speed of a simulation run. Its efficiency over piece-by-piece methods results from the reduced number of events that are observed by the simulator, namely a machine fails or recovers, and a buffer fills or empties. The system evolution between successive events is tracked analytically. Such a model is distinguished as a Class II hybrid model in [30], since it uses simulation and analysis interactively. In [37,38] continuous-flow systems are modeled, whereas discrete-part production lines are examined in [39]. An integrated buffer design algorithm developed in [40] combines PA with the hybrid model to study continuous production lines. Finally in [41], the problem of repair allocation is studied for discrete production lines.

In this article we develop a state space model for production networks based on the discrete-event approach presented in [38,39]. The model describes accurately production lines in both transient and steady-state operations, and is approximate but quite fast for multiproduct networks. The corresponding algorithms are then used to solve two optimization problems. The first concerns the optimal allocation of repair resources and buffer space in production lines . The problem is to maximize the average profit subject to operational constraints. The optimal solution is obtained using a steepest ascent procedure; the gradient information is extracted from simulation experiments by applying a set of simple PA rules. The second problem involves optimal control of production networks with limited number of repairmen. Here we use the model to compare the performance of various state-independent control policies.

II. PRODUCTION LINES

A. SYSTEM DESCRIPTION

A production line is a series arrangement of machines and intermediate buffers, (see Fig.1). Parts enter the first machine and are processed and transported downstream, until they finally leave the system. Machines produce at deterministic but not necessarily equal rates and fail and are repaired randomly. The system is maintained by a number of repairmen. When a machine breaks down a repairman is sent to it immediately. If all repairmen are busy the machine remains inoperable and waits until a repairman is available.

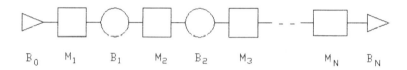

B_0 M_1 B_1 M_2 B_2 M_3 M_N B_N

Fig. 1. A production line with N machines.

The operation of the production line is ruled by the following:

(1) The line consists of N machines and $N+1$ buffers. There is one buffer B_0 at the beginning of the line with finite capacity and another B_N at the end with unlimited capacity.

(2) The nominal production rate of each machine M_i is deterministic. Uptimes and repair times are assumed, for convenience, to be exponential random variables although any type of distribution may be considered. The system is maintained by Q repairmen, where $Q \leq N$.

(3) In each machine there is space for a single workpart. A machine M_i is starved if it has no part to work on and the inventory of the upstream buffer B_{i-1} has been exhausted. Also M_i is blocked if it is prevented from releasing a finished part downstream because B_i is full. In the

sequel the terms empty or starved and full or blocked will be used equivalently.

(4) Starved or blocked machines remain forced down until a workpart or a unit space is available. During these idle periods machines do not deteriorate.

(5) Transportation time of workparts to and from buffers is negligible or is incorporated in the processing time.

It is crucial in our simulation model that production rates be deterministic and known. The assumption that idle machines are not subject to failures has been verified for most production systems [42]. This implies that a machine breaks down only after an amount of production. Time-dependent failures caused by exogenous factors, such as power supply failures, may be easily taken into account by using appropriate random numbers which mark the time of occurence of these phenomena.

The model proposed is event-driven. There are three buffer states: empty, intermediate (partially full), and full. There are three machine states: up, down, and under repair, that is, a repairman is working on a broken machine.

State transitions of machines or buffers will be referred to as **discrete events** or simply **events.**

Now we define the system parameters:

NR_i: nominal production rate (workparts/time-unit) of machine M_i, i=1,...,N.

$1/p_i$:   mean time-to-failure of M_i operating at nominal rate.

$1/r_i$:   mean time-to-repair of M_i.

BC_j: capacity of buffer B_j, j=0,1,...,N.

For any k = 1,2,..., let t_k be the time and e_k the type of the k-th event of the system. The states of the system components (machines and buffers) are evaluated after the k-th event occurence. At time t_k, right after the occurence of e_k the state of the production line is described by:

$$M_i(k) = \begin{cases} U, & \text{if machine } M_i \text{ is up} \\ D, & \text{if } M_i \text{ is down} \\ R, & \text{if } M_i \text{ is under repair} \end{cases}$$

$$B_j(k) = \begin{cases} E, & \text{if buffer } B_j \text{ is empty and } M_{j+1} \text{ is starved} \\ F, & \text{if } B_j \text{ is full and } M_j \text{ is blocked} \\ I, & \text{otherwise an intermediate state.} \end{cases}$$

The above discrete states remain unchanged in the intervals $[t_k, t_{k+1})$ between successive events. We write $B_j(k) = B_j(k+1^-)$ and $M_i(k) = M_i(k+1^-)$.

Next event indicators $\varepsilon_k(c)$ and the associated next-event times $\tau_k(c)$ mark those events that will take place in every component c of the line at time t_k. The next event $\varepsilon_k(M_i)$ for machine M_i belongs in the set {machine-fails, machine-begins-repair, machine-completes-repair}; the next event $\varepsilon_k(B_j)$ for buffer B_j belongs in the set {buffer-empties, buffer-fills, buffer-becomes-not-full, buffer-becomes-not-empty}. Next events and event times are functions of the discrete states. Discrete states, however, do not provide adequate information to predict the evolution of the system. For example, consider a buffer in intermediate state. Its level increases or decreases as workparts arrive or depart until a buffer-fills or buffer-empties event takes place. To compute the type and time of next event we must know the frequencies of part arrivals and departures which in turn are tied to the operations of the upstream and downstream machines.

The dynamic system we study has a more frequently varying section which determines its short-term behavior. The corresponding states will be henceforth referred to as continuous states, in constrast to the slow varying discrete states. At time t_k the continuous state vector consists of:

$BL_j(k)$ level of buffer B_j

$a_j(k)$ residual time-to-next arrival to B_j

$d_j(k)$ residual time-to-next departure from B_j

$TP_i(k)$ total production of machine M_i

$PF_i(k)$ residual parts-to-failure of M_i

$TR_i(k)$ residual time-to-repair of M_i,

for i = 1, 2, ..., N and j = 0, 1, ..., N. In the sequel we drop k whenever this is convenient. The variables $PF_i(k)$ and $TR_i(k)$ are nonincreasing with time. When PF_i equals zero and machine M_i receives an additional workpart, then a breakdown takes place. Similarly when TR_i equals zero the machine resumes operation instantly. In our model these variables are the outputs of appropriate random number generators.

B. SUMMARY OF THE METHOD

A simulation experiment provides a possible evolution of the physical system during a period of operation. We propose a hybrid simulation-analytic approach to the analysis of the DEDS herein which uses simulation and analysis interactively. The algorithm avoids the excessive computational burden of piece-by-piece models by updating the system state only at times when an event occurs, rather than at every part arrival or departure from buffers. Keys to this reduction are the deterministic production rates of machines and the localized effects of event occurrences in the system states.

We call $R_i(k)$ the rate of machine M_i which is a discrete state variable updated after k-th event occurs according to the following rules (see Fig. 2):

1) If a machine is up and it is neither starved nor blocked, then it produces at its nominal rate. The production cycle is the inverse of the production rate.

2) The production cycle of a starved machine M_{i+1} is the time between successive arrivals. Its production rate is set equal to the minimum of the rates NR_{i+1} and R_i.

3) The production cycle of a blocked machine M_i is the time between two successive departures and similarly its rate is set equal to the minimum of the rates of NR_i and R_{i+1}.

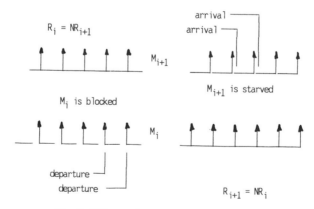

Fig. 2. Effects of events to production rates.

When an event takes place it causes a local change in the discrete state vector. If buffer B_i fills, then its state switches from intermediate to full and the production rate of M_i decreases (see Fig.2.). This may cause chain effects to the components upstream of M_i starting from buffer B_{i-1}. Indeed the output rate of B_{i-1} is reduced to the new production rate of M_i and the buffer will tend to fill. If, however, B_{i-1} was already full, then the production rate of M_{i-1} will be reduced too. Similar changes take place upstream and the phenomenon propagates until a not-full buffer is encountered. When a buffer empties, dual chain effects propagate downstream.

We decompose the chain effect phenomenon using a simple recurrence: at time t_k a buffer-B_i-fills (buffer-B_i-empties) event changes the state variables of the upstream machine M_i (downstream M_{i+1}). We then predict the next event of that machine and of the upstream buffer B_{i-1} (downstream B_{i+1}), by allowing for the possibility that the new buffer-fills (buffer-empties) event be scheduled instantly at time $\tau_k(B_{i-1}) = t_k$. The process then repeats by executing this $(k+1)$-th event. When a machine breaks down, its upstream buffer tends to fill while the downstream one tends to empty. In this case we have two event chains and decompose them similarly.

If $X(k)$ (or $X(k^-)$) is the state vector right after (right before) k-th event occurs, then the proposed simulator performs the following tasks for appropriate functions Γ, Θ, and Π.

1. Computes the time-of-next event in the system

$$t_k = \min \tau_{k-1}(c). \tag{1}$$

2. Updates the state variables right before the execution of k-th event

$$X(k^-) = \Gamma(X(k-1), t_{k-1}, t_k). \tag{2}$$

3. Adjusts the state variables affected by the k-th event

$$X(k) = \Theta(X(k^-), \xi(k)), \tag{3}$$

   where $\xi(k)$ is a random input.

4. Schedules the next events of the affected components c

$$\tau_k(c) = t_k + \Pi(X(k)). \tag{4}$$

In the following sections Eqs. (1)-(4) are developed in detail.

C. EVENT SCHEDULING

At time t right after the occurrence of an event in the system we use analysis to schedule the next events of the components whose state variables were affected by the k-th event.

1. Buffer-Fills

A buffer B_i will become full if M_i produces at a faster rate than M_{i+1} for a period of adequate duration. Note that a necessary condition for this event is that machine M_i becomes blocked and not just that the level of buffer B_i equals its capacity. Also, by convention, the buffer-fills event is encountered when the first part to be blocked arrives at M_i.

Two typical scenarios of this event are illustrated in Figs. 3a, 3b. The continuous line represents a machine operation on a workpart and the arrows represent arrivals to the succeeding buffer. Blank intervals indicate idle periods due to blockage or starvation of machines, and the wavy line denotes a machine under repair.

We examine two different situations which result into a buffer-full event.

Case a. The time-to-next departure from B_i, denoted d_i, is long and the free space of B_i is gradually occupied by parts transmitted from M_i. The condition for this is easily derived by inspection of Fig. 4a:

$$R_i(d_i - a_i) > BC_i - BL(i) . \tag{5}$$

By convention of the buffer-fills event, the event scheduling equations follow directly

$$\tau(B_i) = \begin{cases} t , & \text{if } BL(i) = BC_i \\ t + a_i + \dfrac{BC_i - BL_i - 1}{R_i} , & \text{otherwise} \end{cases}$$

and the production of M_i and M_{i+1}

$$N_i = BC_i - BL_i ,$$
$$N_{i+1} = 0 . \tag{6}$$

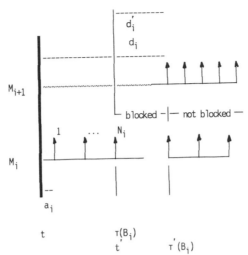

Fig. 3a. Buffer B_i becomes full before next departure.

Case b. Buffer B_i has enough space to accept parts produced by M_i during the transient time d_i, but since M_i produces at a faster rate than M_{i+1} (see Fig. 3b) buffer B_i will eventually fill. The condition for this scenario is

$$R_i > R_{i+1} \qquad\qquad (7)$$

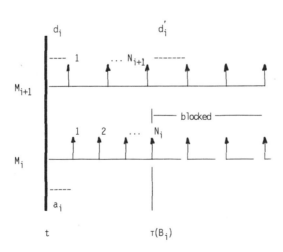

Fig. 3b. Buffer B_i becomes full after a number of departures.

In this case the following holds:

Proposition 1. The production N_i of machine M_i until the buffer-fills event is given by

$$N_i = 1 + INT\{[BC_i - BL_i + R_{i+1}(a_i - d_i)]\, R_i/(R_i - R_{i+1})\} \qquad (8)$$

where $INT(x)$ is the integer part of x.

Proof. Observe in Fig. 3b that the time interval d_i between a departure from M_i and the end-of-processing of the first blocked part lies in an interdeparture interval of M_{i+1}. This implies that

$$\frac{1}{R_i} < d_i' \leq \frac{1}{R_{i+1}}. \qquad (9)$$

From the figure we see that the number of parts produced by M_i after $t + a_i$ is $N_i - 1$. By inspection we obtain the following:

$$N_i - N_{i+1} = BC_i - BL_i \qquad (10)$$

$$N_i = 1 + (\tau(B_i) - t - a_i)\, R_i \qquad (11)$$

$$N_{i+1} = (\tau(B_i) - t - d_i + d_i')\, R_{i+1}. \qquad (12)$$

Upon combining Eqs. (10)-(12) and by eliminating terms $\tau(B_i) - t$ and N_{i+1} we have

$$d_i' = (d_i - a_i) + N_i \left(\frac{1}{R_{i+1}} - \frac{1}{R_i}\right) + \frac{BL_i - BC_i}{R_{i+1}} + \frac{1}{R_i}.$$

By substitution into Eq. (9) we obtain

$$(a_i - d_i) + \frac{BC_i - BL_i}{R_{i+1}} < N_i \left(\frac{1}{R_{i+1}} - \frac{1}{R_i}\right)$$

$$\leq (a_i - d_i) + \frac{BC_i - BL_i}{R_{i+1}} + \frac{1}{R_{i+1}} - \frac{1}{R_i},$$

and (8) follows directly.   ∎

Using Eqs. (8), (11) we can compute the time $\tau(B_i)$ of the buffer-fills event.

2. Buffer-Empties

This event is encountered at time $\tau(B_i)$ when buffer B_i is exhausted and its succeeding machine M_{i+1} has just transmitted a workpart downstream (see Figs. 4a,b). This phenomenon is dual to the blockage and analogous results will be derived.

Case a. If the time-to-next arrival at B_i, denoted a_i, is long enough, then the buffer content is exhausted. By inspection of Fig. 4a the following condition must hold:

$$R_{i+1} (a_i - d_i) > BL_i . \tag{13}$$

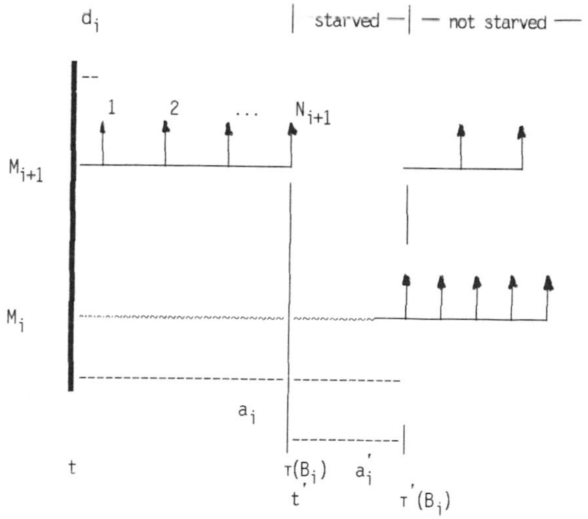

Fig. 4a. Buffer empties before the next part arrival.

The time-to-next event is

$$\tau(B_i) = t + d_i + BL_i/R_{i+1},$$

and the production of M_i and M_{i+1}

$$N_i = 0,$$
$$N_{i+1} = BL_i.$$ (14)

Case b. The buffer has enough parts for the transient period a_i, but machine M_{i+1} produces faster than M_i (see Fig. 4b) and finally B_i becomes empty. The condition for this event is

$$R_i < R_{i+1}.$$ (15)

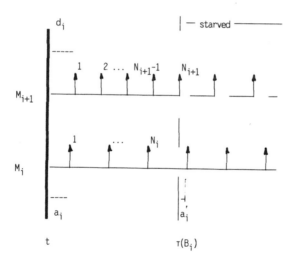

Fig. 4b. Buffer becomes empty after a number of part arrivals.

Then the following proposition holds:

Proposition 2. The production N_{i+1} of machine M_{i+1} until the buffer-empties event is given by:

$$N_{i+1} = 2 + INT\{ [BL_i + R_i(d_i - a_i)] R_{i+1}/(R_{i+1} - R_i) \}. \tag{16}$$

Proof. Observe in Fig. 4b that the interval a_i' lies into the interdeparture interval of M_{i+1} just before occurrence of the buffer-empties event, that is

$$0 < a_i' \le \frac{1}{R_i} - \frac{1}{R_{i+1}}. \tag{17}$$

At time t a workpart is located in M_i and thus the parts-to-next event of M_i and M_{i+1} satisfy

$$N_{i+1} - N_i = BL_i + 1, \tag{18}$$

$$N_{i+1} = 1 + (\tau(B_i) - t - d_i) R_{i+1}, \tag{19}$$

$$N_i = (\tau(B_i) - t - a_i + a_i') R_i. \tag{20}$$

Working as in C1, Case b we obtain Eq. (16).   ∎
The time-of-event $\tau(B_i)$ is computed using Eqs. (16), (19).

3.   Buffer-Becomes-Not-Full and Not-Empty.

If a blocked machine M_i later breaks down or becomes starved, then the buffer-becomes-not-full event occurs in B_i. It is convenient to schedule this event at the departure instant $\tau(B_i)$ of the last blocked workpiece form M_i (see Fig. 3a). We then have the condition:

$$(B_i = Full) \text{ and } (R_i < R_{i+1}) \tag{21}$$

and the time of the not-full event is

$$\tau'(b_i) = t' + d_i'. \tag{22}$$

Similarly we examine the dual system (Fig. 4a). Now M_i is failed and M_{i+1}, although slower, is starved at time t'. The not-empty event is encountered when the operation of M_i is restored and a workpart arrives at M_{i+1}. The condition now is:

$$(B_i = \text{Empty}) \text{ and } (R_i > R_{i+1}), \tag{23}$$

$$\text{and } \tau'(B_i) = t' + a_i'. \tag{24}$$

In the next paragraph we examine how empty and full buffers interact and generate not-full and not-empty events, thus cancelling each other.

4. Event Scheduling of Starved-and-Blocked Machines

In this case machine M_i is located between an empty and a full buffer. It is then forced to wait until a part arrives from the upstream segment and upon completion of processing the part is blocked until an empty space is available in the downstream buffer. This alternation between blocked and starved states occurs frequently when a machine produces at a faster rate than its adjacent ones.

We consider the segment of machines M_{i-1}, M_i, and M_{i+1}, and buffers B_{i-1} and B_i. Let t be the time when the starved-and-blocked event is encountered. We want to find the time τ of next event in the segment. We distinguish the following situations:

1. A starved-and-blocked state is cancelled right after its generation, because either B_i becomes not full or B_{i-1} not empty (see Cases a and b below).

2. A starved-and-blocked state is cancelled after a number of parts have been produced, i.e., outside the transient time of the starved-and-blocked event, again because either B_i becomes not full or B_{i-1} not empty (see Cases c and d).

3. The rates of M_{i-1} and M_{i+1} are equal for a period of time and machine M_i remains starved-and-blocked (see Case e).

Case a: By inspection of Fig. 5a, machine M_i is blocked and its rate has been set equal to the rate of M_{i+1}. However M_{i-1} produces at a slower rate and buffer B_{i-1} is exhausted at time t. After an elapsed time d_i a unit space is available in B_i. The delay time a_{i-1} for the arrival of a new part in M_i is long enough and at the end of its processing cycle machine M_i is no longer blocked. Clearly

$$d_i < a_{i-1} + 1/NR_i. \tag{25}$$

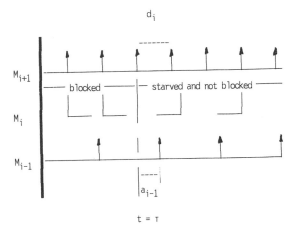

Fig 5a. Cancellation of a starved-and-blocked state
due to an early not-full event.

Here the empty event cancels the blockage directly and thus a not-full event
for B_i occurs simultaneously, i.e.,

$$\tau = t .\tag{26}$$

Case b. A starved machine M_i is slowed down but it fills its downstream
buffer B_i because M_{i+1} produces at a slower rate (see Fig. 5b). Machine M_i
is then blocked and releases the next workpart after time d_i when a unit space
is available in B_i. However, in the meantime M_{i-1} has completed a workpart
and M_i is no longer starved. The condition now is

$$d_i > a_{i-1} + 1/R_{i-1} .\tag{27}$$

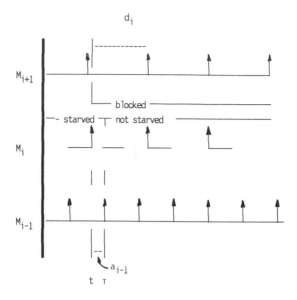

Fig 5b. Cancellation of a starved-and-blocked state
due to an early not-empty event.

A not-empty event for B_{i-1} is scheduled at time

$$\tau = t + a_{i-1} \ .$$

$$(28)$$

Case c. Machine M_{i-1} is slower than M_{i+1}. This situation is depicted in Fig. 5c. Buffer B_i is scheduled to switch from full to an intermediate state.

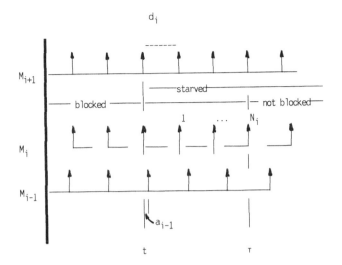

Fig. 5c. Cancellation of a starved-and-blocked state
by a not-full event.

The not-full event occurs upon departure of the last blocked part from M_i.
We then have the following:

Proposition 3. The production N_i of machine M_i until the time-of-next
event is given by

$$N_i = 1 + \text{INT} \left\{ \frac{d_i - a_{i-1} - 1/NR_i}{1/R_{i-1} - 1/R_{i+1}} \right\}. \tag{29}$$

Proof. Observe in Fig. 5c that the end-of-processing time of the $N_i + 1$-th
workpart in M_i is greater than the time at which a single space for this part
is available in B_i. The opposite holds for the first N_i workparts.
This observation directly leads to

$$t + d_i + \frac{N_i}{R_{i+1}} < t + a_{i-1} + \frac{N_i}{R_{i-1}} + \frac{1}{NR_i}, \tag{30}$$

and

$$\tau = t + d_i + \frac{-1 + N_i}{R_{i+1}} \geq t + a_{i-1} + \frac{-1 + N_i}{R_{i-1}} + \frac{1}{NR_i}. \tag{31}$$

Upon combining these inequalities, Eq. (29) results easily. ∎

Using Eq. (20) and the left-hand side of Eq. (31) we obtain the time of next event τ for B_i.

Case d. Now machine M_{i-1} produces at a faster rate than M_{i+1}. Machine M_i is starved, but still faster than M_{i+1}. Thus it fills its upstream buffer B_i (see Fig. 5d). Then the rate of M_i is reduced to the rate of M_{i+1}.

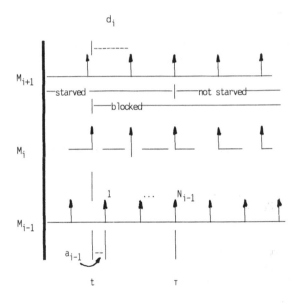

Fig. 5d. Cancellation of a starved-and-blocked state by a not-full event.

This situation is dual to Case c. After machine M_{i-1} processes N_{i-1} parts, a not-empty event will occur and the following is true:

Proposition 4. The production N_{i-1} of M_{i-1} until buffer B_{i-1} becomes not empty is given by

$$N_{i-1} = 1 + \text{INT} \left\{ \frac{a_{i-1} - d_i + 1/R_{i+1}}{1/R_{i+1} - 1/R_{i-1}} \right\}. \tag{32}$$

Proof. In Fig. 5d we observe that the time of arrival of the $1+N_{i-1}$-th workpart at buffer B_{i-1} is less than the time machine M_i is ready to receive it. The opposite is true for the first N_{i-1} workparts.

This observation leads to

$$t + d_i + \frac{-1+N_{i-1}}{R_{i+1}} > t + a_{i-1} + \frac{N_{i-1}}{R_{i-1}}, \tag{33}$$

and

$$t + d_i + \frac{-2+N_{i-1}}{R_{i+1}} \leq t + a_{i-1} + \frac{-1+N_{i-1}}{R_{i-1}} = \tau. \tag{34}$$

In a dual fashion to Case c we obtain Eq. (32). ∎

Using Eq. (32) and the right-hand side of (34) we compute the time-of-next event τ for B_{i-1}.

Case e: If the rates of M_{i-1} and M_{i+1} are equal and Eqs. (25) and (27) are not true, then the times-of-next events in buffers B_{i-1} and B_i are

$$\tau(B_{i-1}) = \tau(B_i) = \infty. \tag{35}$$

The last relation is obtained from Eqs. (34) and (32) by taking the limit as $R_{i-1} \to R_{i+1}$. The machine will remain starved-and-blocked until the end of the production period unless a disturbance is observed, e.g., any of the three machines breaks down, M_{i-1} becomes starved, or M_{i+1} becomes blocked.

5. Machine Event Scheduling

The time between failures depends on the number of parts being processed by a machine and consequently on its production rate. Thus every time the rate is changed, the production horizon has to be modified.

If M_i is up, the time-until-failure is computed by the parts-to-failure PF_i,

$$\tau(M_i) = t + a_i + \frac{PF_i - 1}{R_i} \qquad (36)$$

since the part currently being processed is included in PF_i and the machine will spend a_i time units before it releases it downstream. If M_i breaks down at time t and there is no repairman available, then the time-of-next event at M_i (machine begins repair) is unpredictable and we set $\tau(M_i) = \infty$. If at some time t' another machine completes repair, a repairman becomes available. Therefore we set $\tau(M_i) = t'$, that is machine M_i begins repair immediately. The time-to-repair is computed by invoking the random number generator; the time-of-next event is

$$\tau(M_i) = t' + TR_i. \qquad (37)$$

D. UPDATE EQUATIONS

The update procedure computes the new values of the continuous state variables at instant t_{k-} right before the execution of an event. Our model updates the states of all M_i and B_i in the same routine. The pairs (M_i, B_i) will be referred to as fundamental segments S_i.

Let $q(i)$ denote the index of the last event which caused a change in the states of S_i. Then, dropping i, t_q denotes the time of that event. Also let

n_1 : number of arrivals at B_i in the interval (t_q, t_k),

n_2 : number of departures from B_i during the same period.

We now refer to Fig. 4b and set t_q instead of t, and t_k instead of $\tau(B_i)$. Define

$$A_1 = (t_k - t_q - a_i) R_i ,$$

$$A_2 = (t_k - t_q - d_i) R_{i+1}. \qquad (38)$$

Then the values of n_1 and n_2 are

$$n_m = \begin{cases} 1 + INT(A_m), & \text{if } A_m \geq 0 \\ 0, & \text{otherwise}, \end{cases} \qquad (39)$$

$m = 1,2$, because the elapsed time from the last event may be less than the residual time for an arrival or departure from B_i. We compute total production, parts-to-failure of machine M_i, and new level of buffer B_i:

$$TP_i(k) = TP_i(q) + n_1 \tag{40}$$

$$PF_i(k) = PF_i(q) - n_1 \tag{41}$$

$$BL_i(k) = BL_i(q) + n_1 - n_2 . \tag{42}$$

Finally the update equations for the residual times-to-arrival and departure are

$$a_i(k^-) = \begin{cases} \dfrac{n_1 - A_1}{R_i}, & \text{if } A_1 \geq 0 \\ a_i - (t_k - t_q), & \text{otherwise}, \end{cases} \tag{43a}$$

$$d_i(k^-) = \begin{cases} \dfrac{n_2 - A_2}{R_{i+1}}, & \text{if } A_2 \geq 0 \\ d_i - (t_k - t_q), & \text{otherwise}. \end{cases} \tag{43b}$$

The "minus" superscript is used because the update procedure is activated before the execution of event k, and, as we show below, the above quantities are subject to changes. Finally we set $q(i) = k$ to mark the most recent event upon which we have updated the continuous states of segment S_i.

Computation of Eqs. (38), (39), and (43) must be carried out for both values of m. However if buffer B_i is full or empty and, in addition, $a_i = d_i$ and $R_i = R_{i+1}$, then the updated quantities become

$$n_1 = n_2,$$
$$a_i(k^-) = d_i(k^-), \tag{44}$$

and the computation burden is reduced.

E. EFFECTS OF EVENTS IN STATE VARIABLES

Here we describe the changes of state variables due to event occurrences expressed in Eq. (3). We examine two distinct cases concerning machine and buffer events.

1. Effects of Machine Breakdowns and Repairs

Suppose that at time t_k machine M_i breaks down upon arrival of a workpart. If a repairman is not available, it is not possible to predict the time of-repair of M_i. The apparent departure and arrival times are

$$d_{i-1}(k) = a_i(k) = \infty . \tag{45}$$

If after M_i fails a number of upstream buffers B_{i-1}, B_{i-2}, etc become full and a number of downstream ones B_i, B_{i+1}, etc become empty, their arrival and departure times are also set equal to ∞.

When a repairman is available for M_i the event machine-starts-repair is scheduled immediately. The arrival and departure times become

$$d_{i-1}(k) = a_i(k) = TR_i(k) + 1/NR_i; \tag{46}$$

and the time-to-repair is computed from a random number generator:

$$TR_i(k) = -\frac{\ln\zeta}{r_i} . \tag{47}$$

where ζ is a uniformly distributed random variable in $(0, 1)$. A search procedure traces and updates the full upstream and empty downstream buffers and modifies their delay times according to Eq. (46). Note that these times have been set equal to ∞ by Eq. (45). This search is terminated when a buffer of intermediate state is encountered and the event scheduling routine is invoked for each updated buffer. When M_i is repaired, its repairman is assigned to another failed machine.

Also from the random number generator we compute the new parts-to failure of M_i

$$PF_i(k) = INT \{-\frac{NR_i}{p_i} \ln\zeta\} . \tag{48}$$

TR_i and PF_i are noisy inputs to the system and are expressed by the vector

ξ in Eq. (3).

2. Effects of Events in Buffers

When buffer B_i fills, its residual arrival and departure times are equal and the blocked machine M_i is forced to produce at a slower rate:

$$a_i = d_i,$$
$$R_i = \min(R_i, R_{i+1}). \tag{49}$$

Also if B_{i-1} is not empty, then $d_{i-1} = a_i$.
In a dual fashion we proceed with the buffer empty event

$$d_i = a_i,$$
$$R_i = \min(R_i, R_{i+1}), \tag{50}$$

and if B_{i+1}, then $a_{i+1} = d_i + 1/NR_{i+1}$.

In the above two cases, if machine M_i becomes starved-and-blocked, the next event scheduling is performed according to Section II, C, 4.
If buffer B_i becomes not-full, then we set

$$a_i = d_{i-1}. \tag{51}$$

If machine M_i is starved, then we set $R_i = \min(NR_i, R_{i-1})$, otherwise $R_i = NR_i$.
In a dual fashion we proceed with the not-empty event

$$d_i = a_{i+1}. \tag{52}$$

If machine M_{i+1} is blocked we set $R_i = \min(NR_i, R_{i+1})$, otherwise $R_i = NR_i$.

3. Event Priority Rules - The Algorithm

Since machine failures and all buffer related events occur upon departure or arrival of workparts to buffers, the probability of encountering two simultaneous events conditioned on a single move is not negligible. In this case one event dominates the other. We examine two such cases.

a. A machine breaks on a workpart which in turn is going to be blocked; then the failure event must be executed first. The production of workparts may be delayed for a long enough period to cancel the blocking event.
b. Machine M_i is scheduled to be starved-and-not-blocked simultaneously. Then the buffer empty event is executed first according to Section II, C, 4.

The discrete-event algorithm is as follows:

1. Initialize the line. Set total simulation time and number of raw parts in B_0, initial buffer levels and capacities, nominal production rates and residual arrival and departure times of buffers, machine mean times-to-failure and repair, length N of the production line. Trace the line downstream and schedule next events.
2. Determine the next event. Find those events with the smallest time and select one which complies with the priority rules. If the time-of-next event does not exceed the total simulation time, execute the appropriate event and return to (2), otherwise terminate the simulation.
3. Terminate the simulation. Trace the line downstream and update all system variables.

To execute events we apply Eqs. (2)-(4) and compute the state variables of appropriate fundamental segments:

1. Segments S_i, S_{i+1} when buffer B_i becomes empty or not empty.
2. Segments S_i, S_{i-1} when any other event takes place in M_i or B_i.

D. APPLICATIONS

The algorithm above has been developed into a FORTRAN 77 code. To evaluate its computational efficiency we also wrote a piece-by-piece simulator as a benchmark. The two algorithms are compared using common random numbers.

The computational requirements of piece-by-piece models are almost proportional to the lenght N of the line and the total amount of production during a simulation period. The runtime of our hybrid model depends on the number of events. Although this number is smaller than the number of departures and arrivals of workparts at every machine, their execution requires significant amounts of computation due to the complexity of Eqs (2) and (4). However the algorithm is expected to be fast if machine failures and events in buffers are rare. The relative speed of the event-driven (e-d) model as compared to that of the piece-by-piece (p-p) simulator is defined by the ratio

CPU time of p-p simulator

———————————————————————

CPU time of e-d model

and is shown in Fig. 6 as a function of failure rates and buffer capacities. As an example we considered a line of 40 machines and 40 repairmen with repair rates 1.0. The production rates were $NR_{5i} = 10$, $NR_{5i+1} = 8$, $NR_{5i+2} = 12$, $NR_{5i+3} = 15$, $NR_{5i+4} = 16$, for $i = 0,1,...,8$.

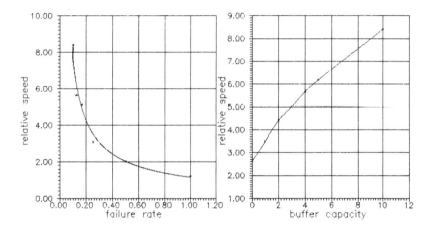

Fig. 6. Relative speed vs failure rates
and buffer capacities.

Since in most actual cases machines are relatively reliable, it turns out that our model is a practical tool for the analysis of production lines, in both transient and steady-state operation. In the rest of this section we develop an efficient procedure for optimal design and control of production lines.

1. Problem Formulation

Let $r = \{r_1, r_2, ..., r_N\}$ be the set of repair rates and α the total amount of repair rate to be distributed among machines. Also $BC = \{BC_1, ..., BC_{N-1}\}$ is the set of buffer capacities, β the total buffer space available, and $TH(r;BC)$ the throughput rate of the line corresponding to r and BC. Then our problem is formulated as follows (see also [31]):

$$\text{maximize } TH(r;BC) \tag{53}$$

subject to constraints

$$\sum_{i=1}^{N} r_i = \alpha ,$$

$$\sum_{j=1}^{N-1} BC_j = \beta . \tag{54}$$

We define $\partial TH/\partial r_i = S(r_i)$ and $\partial TH/\partial BC_j = S(BC_j)$. We assume that the buffer capacities are large enough so that the following approximation is acceptable

$$S(BC_j) \approx TH(BC_j + 1) - TH(BC_j) . \tag{55}$$

Then the necessary conditions for (53), (54) are

$$S(r_i^*) - \lambda_1 = 0 , \quad i = 1, ..., N$$

$$S(BC_j^*) - \lambda_2 = 0 , \quad j = 1, ..., N-1 \tag{56}$$

and r_i^*, BC_j^* are the optimal repair rates and buffer capacities. A gradient scheme for the sequential approximation of the optimal point is

$$r_i(k+1) = r_i(k) + J_k \left[S(r_i(k)) - (1/N) \sum_{m=1}^{N} S(r_m(k))\right], \tag{57a}$$

$$BC_j(k+1) = BC_j(k) + J_k [S(BC_j(k)) - (1/(N-1)) \sum_{m=1}^{N-1} S(BC_m(k))], \quad (57b)$$

where J_k is an acceleration parameter determined empirically. Convergence is a crucial issue for the algorithm. Although convexity of the throughput rate has not been proved for the general case, small system analytic solutions do not indicate the existence of multiple local optimum points to one of which the sequential procedure might be trapped (see also [40]).

Next we describe an efficient method for obtaining partial derivatives of the throughput with respect to buffer capacities and repair rates.

2. Perturbation Analysis

The basis of perturbation analysis is a set of simple rules which describe the effect of a change in a system parameter on the average throughput rate. The technique leads to a direct computation of gradients with respect to various design parameters using a single simulation run. As an example, an increase δ in the mean-time-to-repair $1/r_i$ of M_i, causes a delay d in the operation of M_i which is proportional to δ (see Eq. (47)). This delay may be translated to adjacent machines according to certain propagation rules or even cancelled. Upon termination of the simulation, the total effect on the production time is given by the sum of those d's which are realized by the last machine.

The perturbation propagation rules are summarized below [45]:

1. When a machine M_i is blocked, all perturbations currently residing at M_{i+1} are copied by M_i.
2. In a dual fashion if M_i is starved, perturbations accumulated at M_{i-1} are copied by M_i.

Let

SP   : simulation period

$s(i,j)$: total delay in production time of M_i caused by a decrease of the repair rate of M_j to the value of $r_j/2$. Here we divide r_j by 2 for convenience. As will become clear below, the computation of gradients is not affected by this division.

Partial derivatives with respect to repair rates are obtained by the following algorithm.

1. Initialize the line and set $s(i,j)=0 \; \forall \, i,j$. When an event occurs execute one of the following:

2.   When M_i is repaired, then $s(i,i) = s(i,i) + TR_i$, where TR_i is computed from (47).
3.   If B_i becomes full, $s(i,j) = s(i+1,j)$ $j=1,...,N$.
4.   If B_i becomes empty, $s(i+1,j) = s(i,j)$ $j=1,...,N$
5.   At time SP terminate the simulation and calculate the gradients

$$\frac{\partial TH}{\partial r_i} = \frac{\partial(TP(N)/SP)}{\partial r_i} = -\frac{TP(N)}{(SP)^2}\frac{\partial SP}{\partial r_i}$$

$$\approx -\frac{TP(N)}{(SP)^2}\frac{s(N,i)}{r_i/2} = S(i) . \tag{58}$$

To obtain throughput rate sensitivities with respect to buffer capacities we use a procedure presented in [40]. Let

$\sigma(i, j)$:   total change in buffer level BL_i caused by a unit increase of BC_j, $i=0,1,2,...,N$ $j=1,2,..., N-1$.

The discrete-event algorithm is extended as follows:

1.   Initialize the line and set $\sigma(i,j) = 0$ $\forall$ i,j.

2.   When an event occurs execute one of the following:

3.   If B_i becomes full, then
     a.   If B_i has visited an empty state recently, then set $\sigma(i-1,i) := \sigma(i-1,i)-1$.
     b.   If B_i has visited a full state recently, then $\sigma(i-1,j) = \sigma(i-1,j) + \sigma(i,j)$, and $\sigma(i,j) = 0$, $\forall$ j.

4.   If B_i becomes empty, then
     a.   If B_i has visited a full state recently, then set $\sigma(i+1,i) = \sigma(i+1,i)+1$
     b.   If B_i has visited an empty state recently, then set $\sigma(i+1,i) = \sigma(i+1,j) + \sigma(i,j)$, and $\sigma(i,j) = 0$, $\forall$ j.

5.   At time SP terminate the simulation and calculate the sensitivities

$$TH(BC_j+1) - TH(BC_j) = \frac{\min\{|\sigma(0,j)|, |\sigma(N,J)|\}}{SP}.$$

The sizes of perturbations in the parameters r_i and BC_j affect crucially the accuracies of gradient estimation. It is assumed that the evolution of the perturbed system resulting from a change in a single parameter, is similar to the evolution of the initial system. Two systems are said to be similar if their scheduled event sequences for both of them are identical although the corresponding event times may differ. A number of experimental and theoretical results ([31], [34], [41], [45]) ensure that even when this assumption is not valid, in most design cases the maximum estimation errors are 10%. In addition, as explained in [45], the estimates $S(r_i)$ are strongly consistent.

3. Example

Consider a four-stage production line with $NR_1 = 10$, $NR_2 = 8$, $NR_3 = 12$, $NR_4 = 15$ and $p_i = 0.1$, $i = 1,...,4$. The parts-to-failure are geometrically distributed with parameters $\exp(-p_i/NR_i)$, and the times-to-repair exponentially distributed with mean values $1/r_i = 1$. The system is maintained by four repairmen, that is, there is always a repairman available for a failed machine. However, we would like to know whether we should favor certain machines by assigning more than one repairman to them, thus speeding up their repair process. We want to maximize the system throughput subject to $\alpha = 4$ and $\beta = 15$.

The total production was set to 50.000 items and the termination condition of the gradient algorithm was $TH(k+1) - TH(k) < 10^{-3}$. The results are summarized in Table I.

Table I. Five iterations of the gradient algorithm

k	BC_1	BC_2	BC_3	r_1	r_2	r_3	r_4	TH(K)
0	5	5	5	1.00	1.00	1.00	1.00	6.690
1	6	6	3	1.01	1.05	1.00	0.94	6.751
2	6	8	1	1.01	1.08	1.00	0.91	6.790
3	6	9	0	1.02	1.10	0.99	0.89	6.806
4	6	9	0	1.02	1.13	0.98	0.87	6.811
5	6	9	0	1.03	1.15	0.97	0.85	6.812

In the beginning, the buffer space and repair resources are equally distributed. At the fifth iteration we observe that the total inventory space is distributed between the buffers surrounding the slowest machine M_2. This happens in order to eliminate blockage or starvation phenomena. Also the repair rate of M_2 should be increased. This is possible in practice by employing partially the repairman of M_4. These results agree with intuition since they suggest that the slowest machine should be favored.

III. PRODUCTION NETWORKS

A. SYSTEM DESCRIPTION

In this chapter we examine production networks of general geometry which can manufacture a variety of part types simultaneously. Our purpose is to develop a hybrid simulation/analytic model by observing the discrete events which take place during the production period. We consider the section of a production network shown in Fig. 7 which consists of workstations W_l, W_j, W_g and intermediate buffers B_{lj}, B_{mj}, ..., B_{jg}. Buffer B_{jg} is located between workstations W_j and W_g, which are called adjacent; also W_j is the upstream and W_g is the downstream workstation with respect to B_{jg}. The production of W_j is distributed to downstream machines through buffers B_{jp}, B_{jg}, ... Raw parts enter the network from an infinite source B_{in} and after being processed they eventually leave the system to an output buffer B_{out}. A workstation W_j consists of a number of parallel machines labelled M_{ij} which may produce at different rates and fail and are repaired randomly. There is a limited number Q of repairmen in the network. Each workstation performs a specific type of operation.

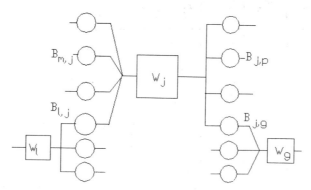

Fig. 7. Section of a production network

There are several part-types. Each one requires a fixed set of operations and follows a certain route in the production system. All types of raw parts are available from the source buffer B_{in} and the output buffer B_{out} can accept all types of finished products. Workstations may perform assembly operations on several part-types to produce composite parts. A composite part is formed from constant numbers of parts from each type called assembly coefficients. For example, W_j needs δ_{lj} parts from W_l, δ_{mj} parts from W_m, etc, to produce a composite part. The flow of a part-type may also split into two or more streams as a result of connecting workstations with several buffers downstream.

These assumptions appear to be restrictive for the study of controlled flexible manufacturing systems (FMS's) in which a workstation performs more than one tasks and various part-types share a single buffer and may have more than a single routing possibility. However, in most cases the levels of each part type in a buffer are not allowed to exceed specified limits, which is analogous to our assumption of multiple downstream buffers of finite capacities. Buffer level control is achieved by switching the production rates of workstations to various modes for each part-type.

Machine loading and splitting of workparts are performed according to a specific protocol. We assume for convenience two possibilities for non-state-dependent routing strategies: the familiar first-come-first-served (FCFS) and the fixed-priority (FP) rule. The latter results in a maximum utilization of high priority machines and downstream buffers of every workstation over those of lower priorities. We could incorporate any flow control law in the model but such an attempt would fall beyond the scope of this article.

As in the case of production lines we consider three machine states: up, down and under repair, and three buffer states: full, empty and intermediate. The corresponding events are: machine completes repair, fails and starts repair, and buffer fills, empties, becomes not-empty and becomes not-full. Next we use an event domain formalism, and the notation of the previous chapter to describe the evolution of the system during a production period.

B. SUMMARY OF THE METHOD

We examine first how various events influence the system evolution. Referring to Fig. 7 we make the following remarks.

1. Suppose that machine M_{ij} of workstation W_j is ready to send a part to buffer B_{jq} which is full. Then a buffer-fills event takes place. The machine will attempt to send the overflowing workpart to other buffers B_{jp} which are partially full. These buffers will tend to fill. Thus the

buffer-fills event generates chain events downstream. However, if all downstream buffers are full, the machine and, therefore, the workstation will be forced to produce at slower rates. Consequently, the upstream buffers B_{mj}, B_{lj}, etc, of the workstation will tend to fill and thus the buffer-fills event will propagate upstream.

2. When buffer B_{lj} becomes empty, it cannot feed the downstream workstation W_j with parts of a certain class. The workstation will not produce any workparts and therefore the inventories of its downstream buffers B_{jp}, B_{jg}, etc., will tend to be exhausted. Thus the buffer-empties event propagates downstream and is dual to the buffer-fills event. In addition, the other upstream buffers B_{mj} of W_j will tend to fill because their parts cannot be assembled due to shortage of parts from B_{mj}.

3. If machine M_{ij} breaks down and the corresponding workstation is neither starved nor blocked, the total production rate of W_j is reduced. The upstream buffers tend to fill and the downstream ones tend to empty.

4. In a dual fashion, if a machine completes repair and becomes operational, it tends to exhaust the contents of its upstream buffers and fill its downstream ones.

In order to find the chain events which take place in the system after the occurrence of the k-th event in addition to the definitions of Chapter II we introduce the following

n_j number of machines of W_j

$R_j(k)$ total production rate of W_j right after the k-th event.

$NR_j(k)$ production rate of W_j if it were neither blocked nor starved

$I_{lj}(k)$ inflow rate to buffer B_{lj},

$O_{lj}(k)$ outflow rate from B_{lj} .

The quantity $NR_j(k)$ will be referred to as the **capacity** of workstation W_j and is equal to the sum of the nominal rates of the operational machines.

The analysis of Section II, C which yielded the event scheduling equations of production lines, was quite complex and its extention to production networks becomes prohibitive due to its combinatorial nature. We shall thus make the assumption of continuous goods flow rather than discrete-part routing in the network. Thus the model is not exact. Yet, its accuracy turns out to be exceptionally good for a wide range of practical applications.

As a consequence the residual times for arrivals and departures from buffers are not necessary for our analysis; if a machine is down or under repair its production rate is zero.

By assumption, splitting, merging and machine loading of workparts depend on the flow control protocol. Under the FCFS rule the production rate of an operating machine M_{ij} and the inflow rate to buffer B_{jg} having intermediate or zero inventory level satisfy

$$R_{ij} = \frac{NR_{ij}}{NR_j} R_j ,$$ (56a)

$$I_{jg} = \frac{R_j}{n}$$ (56b)

where n is the number of not-full buffers which are downstream of W_j. For the FP rule let $M_{[1]j}$, $M_{[2]j}$, ... be the sequence of operating machines of W_j in descending priority, and $B_{j[1]}$, $B_{j[2]}$, ... the corresponding sequence of not full downstream buffers. Then

$$R_{[1]j} = \min \{NR_{[1]j}, R_j\}$$

$$R_{[i]j} = \min \{NR_{[i]j}, R_j - \sum_{s=1}^{i-1} R_{[s]j}\} , \quad i=2, ..., n_j .$$ (57a)

Also for $R = R_j$,

$$I_{j[1]} = R ,$$

$$I_{j[g]} = 0 , \quad g=2, ..., n .$$ (57b)

If machine $M_{[1]j}$ breaks down (buffer $B_{j[1]}$ becomes full) we exclude that component from the corresponding sequence, (set $R = R_j - O_{j[1]}$) and apply Eqs. (57a) (resp. Eqs. (57b)) again.

The proposed hybrid algorithm uses Eqs. (1), (3) (the simulation part) and Eqs. (2), (4) (the analytic part) interactively to mimic the system evolution. In the next sections we develop these equations in explicit forms.

C. EVENT SCHEDULING

We observe the system at time t right after the occurrence of an event. Suppose that this event changed the state variables of $B_{\ell j}$, W_j and B_{jg} shown in Fig. 7. We develop explicit expressions for the event scheduling equations (4).

1. Scheduling Events in Buffers

Let $\tau(B_{jg})$ be the time-of-next event for buffer B_{jg}.

a. Buffer B_{jg} fills. If the input rate of buffer B_{jq} is greater than the output rate, then the buffer will fill in time:

$$\tau(B_{jg}) = t + \frac{BC_{jg} - BL_{jg}}{I_{jg} - O_{jg}}, \qquad (58)$$

for $I_{jg} > O_{jg}$.

b. Buffer B_{jg} empties. If the ouput rate of the buffer is greater than its input rate, then the buffer will empty in time:

$$\tau(B_{jg}) = t + \frac{BL_{jg}}{O_{jg} - I_{jg}}, \qquad (59)$$

for $O_{jg} > I_{jg}$.

c. Constant buffer level. If the buffer rates are equal, then the buffer will stay at the same state, i.e.

$$\tau(B_{jg}) = \infty , \quad \text{for equal rates.} \qquad (60)$$

d. Not-full and not-empty events. We consider buffer B_{jg} which is full because either a machine in the downstream workstation W_g is under repair, or the workstation is starved and forced to produce at a slower rate. The input rate of the buffer is reduced and set equal to the output rate. At time t when the machine is repaired or the workstation is not forced down any more, buffer B_{jg} becomes not full immediately.

In a dual fashion, if buffer B_{lj} is empty and the production rate of W_l is increased at time t, then B_{lj} becomes not empty.

For these cases we have

$$\tau(B_{jg}) = t, \quad \text{for } O_{jg} > I_{jg},$$ (61)

and

$$\tau(B_{lj}) = t, \quad \text{for } I_{lj} > O_{lj}.$$ (62)

A buffer-B_{jg}-becomes-not-full (B_{lj}-becomes-not-empty) event occurs only after a machine is repaired, or after an upstream buffer becomes not full (a downstream buffer becomes not empty).

2. Machine Event Scheduling

The time-to-failure of an operating machine M_{ij} is computed by the number of parts-to-failure which is a variable with known distribution. The time $\tau(M_{ij})$ at which a failure occurs is given by:

$$\tau(M_{ij}) = t + \frac{PF_{ij}}{R_{ij}}.$$ (63)

When a machine is under repair, the time-of-next event is computed by:

$$\tau(M_{ij}) = t + TR_{ij},$$ (64)

where TR_{ij} is the time-to-repair of M_{ij}. This time is a random variable of known distribution.

D. UPDATING AND ADJUSTING THE STATE

To execute the next event we need to update the relevant continuous state variables. Then we proceed by adjusting the discrete states which are affected by the event.

1. Update Equations

Let $q(j) = q$ and k be two successive events which cause a change to the production rate of M_j. At time t_{k^-} right before the execution of the k-th event we update the states of machines M_{ij}, and buffers B_{ll} and B_{jg}.

$$TP_{ij}(k) = TP_{ij}(q) + R_{ij}(q) (t_k - t_q)$$

and $PF_{ij}(k) = PF_{ij}(q) - R_{ij}(q) (t_k - t_q), \qquad$ if M_{ij} is up

or $\quad TR_{ij}(k) = TP_{ij}(q) - (t_k - t_q), \qquad$ if M_{ij} is down.

The new buffer level of B_{lj} becomes

$$BL_{lj}(k) = BL_{lj}(q) + [I_{lj}(q) - O_{lj}(q)] [t_k - t_q] . \qquad (66)$$

Finally we set $q(j) = k$ to mark the new most recent event which affected W_j.

2. Effects of Changes in Machine States

We examine the effects of machine failures and repairs.

a. Machine-is-repaired. Let q be the previous event which took place at W_j. Suppose that at time t_k machine M_{ij} is repaired. Then we mark the new machine state and adjust the capacity of W_j,

$$NR_j(k) = NR_j(q) + NR_{ij}. \qquad (67)$$

The number of parts-to-failure is found from a random number generator

$$PF_{ij}(k) = - \frac{NR_{ij}}{p_{ij}} \ln \zeta , \qquad (68)$$

where p_{ij} is the failure rate of machine and, again, ζ is a uniform random number in $(0, 1)$. If the workstation is not slowed down we set

$$R_{ij}(k) = NR_{ij} . \qquad (69)$$

However if in the interval (t_q, t_k) the workstation was starved or blocked, its total production rate $R_j(k)$ of W_j is unchanged and we compute the rates of machines and downstream buffers using Eqs. (56) or (57). The new rates from the upstream buffers B_{mj} are proportional to the corresponding assembly coefficients

$$O_{mj}(k) = \delta_{mj} R_j(k) . \tag{70}$$

The repairman of M_{ij} is now ready to serve another failed machine of the production network. Thus a machine-begins-repair is scheduled instantly.

b. Machine-fails and machine-starts-repair. Suppose now that at time t_k machine M_{ij} breaks down. Then we have

$$R_{ij}(k) = 0, \tag{71}$$

$$NR_j(k) = NR_j(q)\text{-}NR_{ij}.$$

We now have two possible scenarios. If all repairmen are busy it is not possible to determine the time-to-begin-repair of M_{ij}. We thus set

$$\tau(M_{ij}) = \infty . \tag{72}$$

Otherwise, if a repairman is available the time-to-repair is found by the random number generator

$$TR_{ij}(k) = -\frac{1}{r_{ij}} \ln \zeta \tag{73}$$

and the machine is set under repair.

If the workstation is blocked in the interval $[t_q, t_k)$, then the new machine and downstream-buffer rates are computed from Eqs. (56) or (57).

If the workstation is not blocked, then under FP policy we use Eqs. (57b) to compute the inflow rates to downstream buffers. Under FCFS policy we apply the following iterative procedure: Let F_j be the set of full buffers downstream of W_j, and G_j the set of partially full and empty buffers. The inflow rate to a full buffer must not exceed its output rate, whereas a not full buffer of G_j can accept parts at any rate until t_k. If after the time of failure the inflow rates to some full buffers become less than the output ones, then their states become intermediate instantly. The rate allocation procedure is as follows:

1.   The set and the total number of buffers where we allocate rates are

$D = F_j \cup G_j$ and $L = n$ respectively. The total undisposed rate is $R = R_j$.

2. Set inflow rates of all buffers in D equal to R/n.
3. Trace the buffers in D. If a buffer is full and its outflow rate is smaller than its inflow rate, then set the latter equal to the former, reduce the total undisposed rate by this ammount, exclude this buffer from the set D, and reduce L by one. Go to (2).
4. Stopping rule. If $L > 0$ go to (2), otherwise stop. If there is still an overflow, the production rate of the workstation is reduced immediately.

In the sequel we use the acronym RAP in referring to the rate allocation procedure for the downstream buffers, under either FCFS or FP flow control policies.

Finally, the new outflow rates from the upstream buffers are computed according to Eq. (70).

3. Effects of Changes in Buffer States

We consider the four events: buffer-fills, buffer-becomes-not-full, buffer-empties, and buffer-becomes-not-empty. Again q denotes the previous event which affected the state variables of the buffers.

a. Buffer fills. If buffer B_{jg} fills at time t_k, its input rate is set equal to its output rate:

$$I_{jq}(k) = O_{jq}(k) < I_{jg}(q). \qquad (74)$$

This event has some effects to other buffers which are adjacent to workstation W_j. If W_j can redistribute the excess of production to other (not-full) downstream buffers, then their inflow are computed from Eqs. (56b) or (57b) and there is no further effect to the workstation . If, however, all the downstream buffers are full, then the production rates of the workstation are reduced according to Eqs. (56a), (57a) and

$$R_j(k) = R_j(q) + I_{jg}(k) - I_{jg}(q) . \qquad (75)$$

The effect of this reduction is then instantly realized by the upstream buffers of W_j as

$$O_{lj}(k) = \delta_{lj} R_j(k). \qquad (76)$$

If any of these buffers is full, then a new buffer-fills event is scheduled to take place instantly.

b. *Buffer-becomes-not-full.* We give an example here. Suppose that the previous event q occurred after a machine in workstation W_g broke down and buffer B_{jg} became full. When the machine is repaired, the outflow from B_{jq} increases and a not-full event is scheduled immediately. If workstation W_j is not starved (it is possible for a workstation to be starved only when at least one of its downstream buffers is not full), its production rate is set equal to its available one and the rates of its upstream buffers are adjusted as in Eq. (76). The new inflow rates to the buffers downstream of W_j are found using either of the RAP procedure described in Case b of the previous paragraph.

c. *Buffer-becomes-empty.* If buffer B_{lj} empties at time t_k, its output rate is reduced.

$$O_{lj}(k) = I_{lj}(k) < O_{lj}(q) . \qquad (77)$$

The production rate of workstation W_j as well as the outflow rates from the rest of its upstream buffers B_{mj} are then proportionally reduced. These changes are consequences of the assembly operation. Therefore

$$R_j(k) = O_{lj}(k) / \delta_{lj} ,$$

$$O_{mj}(k) = \delta_{mj} R_j(k). \qquad (78)$$

If any of these buffers had been empty by time t_k, its state will become intermediate immediately due to instant reduction of O_{mj}. Again we apply RAP to adjust the inflows to buffers downstream of W_j, and Eqs. (56b) or (57b) for the new production rates of machines M_{ij}.

d. *Buffer-becomes-not-empty.* In a dual fashion to Case b, a buffer B_{lj} becomes not empty when a machine of workstation W_l is repaired. Then workstation W_j, which was previously starved, can produce at a faster rate which is determined by its capacity and the new inflow rate of B_{lj}:

$$O_{lj}(k) = \min \{I_{lj}(k), NR_j(k) \, \delta_{lj}\} ,$$

$$R_j(k) = O_{lj}(k) / \delta_{lj} . \qquad (79)$$

The outflows from the rest upstream buffers B_{mj} are increased:

$$O_{mj}(k) = \delta_{mj} R_j(k). \qquad (80)$$

As in Case c, we adjust the input rates of the downstream buffers and the production rates of the machines of W_j.

E. APPLICATIONS

Following the steps of section II, E, 3 we have developed the event-driven algorithm in FORTRAN 77 code. In this paragraph we investigate the model's accuracy and speed versus a piece-by-piece simulator in the study of systems with different configurations and parameters, and present a control application for a system with limited number of repairmen.

1. Computational experience

As discussed in the previous chapter, a critical factor for the computational efficiency of our model is the frequency of failures. We examined the two-product system of Fig. 8. The numbers in boxes denote numbers of parallel machines and those in buffers are the corresponding capacities; the assembly coefficients are read along the arcs connecting buffers to downstream workstations. The system operates under FCFS routing policy and a repairman is always available for each machine.

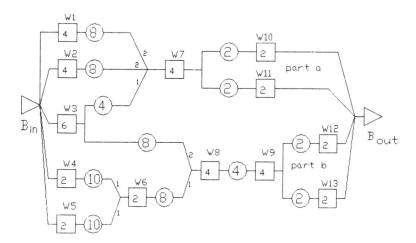

Fig. 8. A two-product network.

The machines' production and repair rates are 10.0 and 1.0 respectively. Figures 9 a,b show how the throughput relative errors and speed of the event-driven model vary as the failure rates increase.

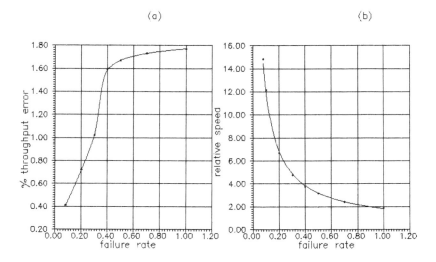

Fig. 9. Performance of the event
driven model vs failure rates

The simulation horizon was set at 3,000 time units and the estimates were obtained from three independent runs. Figure 9a shows the maximum error of the two products which in all cases but one corresponds to the throughput of product a. From Fig. 9 we see that the model performance improves with decreasing frequency of failures.

A second issue of interest is the efficiency of our algorithm for large systems. For this study we analyzed a serial system of N workstations with intermediate buffers having capacities 10. Each workstation consists of three machines with NR=10.0 p=0.05 and r=0.5. Figure 10 shows the relation between relative speed and N. It turns out that our algorithm is by far more efficient than conventional simulators especialy for large systems. Also in this example the throughput errors were less than 0.1% whereas the average buffer level errors were less than 2%.

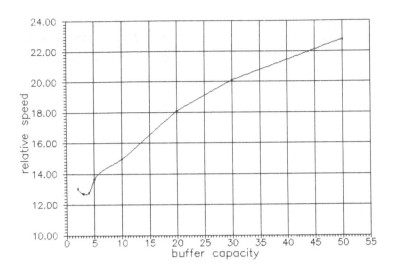

Fig. 10. Efficiency vs system size

2. A control application

Now we study the problem of finding the optimal assignment of a limited number of repairmen to broken machines. We consider non preemptive policies and the objective is to maximize throughput. The only analytical results in the literature concern two-stage systems and production lines with infinite buffer capacities (see [43], [44]). The idea here is to use our model for testing various rules of thumb, such as

First-come-first-served (FCFS)
Shortest (longest)-expected-repair-time (SERT/LERT)
Shortest-expected-up-time (SEUT)
Smallest (largest)-expected-production-to-failure (SEPF/LEPF)
Smallest (greatest)-efficiency-in-isolation (SEI/GEI).

The efficiency in isolation of a machine M_{ij} is its stand-alone average throughput rate $NR_{ij}r_{ij}/(r_{ij}+p_{ij})$.

We considered the single product system of Fig. 11. The results are summarized in Table II for various numbers of repairmen.

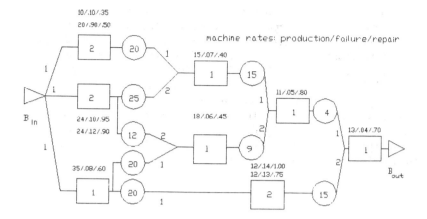

Fig. 11. A single product test system.

Table II. Performance of control policies
for different numbers of repairmen

POLICY	REPAIRMENS, Q				
	1	2	3	4	5
LERT	4.675	5.775	5.900	5.916	5.932
SERT	4.728	5.765	5.914	5.921	5.932
SEUT	4.622	5.755	5.905	5.921	5.932
SEPF	4.574	5.766	5.905	5.921	5.932
LEPF	4.821	5.784	5.897	5.916	5.932
FCFS	4.669	5.769	5.905	5.912	5.932
SEI	4.666	5.777	5.916	5.921	5.932
LEI	4.709	5.780	5.898	5.916	5.982

When the number of repairmen Q is 1 or 2 the machines spend long periods waiting to begin repair and the whole system is slowed down. For values of $Q \geq 5$ the system throughput remains constant. Therefore if a decision were to be taken for Q, its value should not have exceeded 4 or 5. Also, note that LEPF rule performs better for Q = 1,2 whereas SEI should be preferred for $Q \geq 3$. The last policy agrees with intuition suggesting that the slowest on the average workstations, should be favored by the repairmen. From these and other experiments reported in [41] concerning production lines it seems that there is not a unique optimal policy for all systems. At any rate, our algorithm can be applied to determine the optimal policy at low computational cost compared to conventional simulators.

REFERENCES

1. P.J. Burke, "The Output of a Queueing System", *Operations Research* 4(6), pp. 699-704 (1956).
2. J.R. Jackson, "Networks of Waiting Lines," *Operations Research*, 5, pp. 518-521 (1957).
3. W.J. Gordon, and G.F. Newell, "Closed Queueing Systems with Exponential Servers," *Operations Research* 15, pp. 254-265 (1967).
4. F. Baskett, K.M. Chandy, R.R. Muntz, and F.G. Palacios, "Open, Closed and Mixed Networks of Queues with Different Classes of Customers," *Journal of the Association for Computing Machinery* 22, pp. 248-260 (1975).
5. J.A. Buzacott, "Automatic Tranfer Lines with Buffer Stocks," *International Journal of Production Research* 5(3), pp. 183-200 (1967).
6. Y.A. Phillis, "A Mathematical Study of Unreliable Production Systems," *Proceedings of the 16th Conference on Information Sciences and Systems,* Princeton University (1982).
7. G.C. Hunt, "Sequential Arrays of Waiting Lines," *Operations Research* 4(6), pp. 674-683 (1956).
8. G.T. Artamonov, "Productivity of a Two-Instrument Discrete Processing Line in the Presence of Failures," *Cybernetics (English translation)* 12(3), pp. 464-468 (1977).
9. M.F. Neuts, Matrix-Geometric Solutions in Stochastic Models, Johns Hopkins University Press, Baltimore (1981).
10. S.B. Gershwin, and O. Berman "Analysis of Transfer Lines Consisting of Two Unreliable Machines with Random Processing Times and Finite Storage Buffers," *AIIE Transactions,* 13(1), pp. 1-11 (1981).
11. P.P. Bocharov, "Queueing System of Limited Capacity with State Dependent Distributions of Phase Type," *Automation and Remote Control (English translation),* pp. 31-38 (1985).
12. V.S. Kouikoglou, "A Model for Two-Stage Production Systems with Scrapping," *Foundations of Decision and Computing in Engineering* 15(2), pp. 77-93 (1990).
13. F.S. Hillier and Boling, "Finite Queues in Series with Exponential or Erlang Service Times - A Numerical Approach," *Operations Research* 15(2), pp. 651-658 (1967).
14. P.J. Kuhn, "Approximate Analysis of General Queueing Networks by Decomposition," *IEEE Transactions on Communications* C-27(1), pp. 113-126 (1979).
15. Y. Takahashi, H. Miyahara, and T. Hasegawa, "An Approximation Method for Open Restricted Queueing Networks," *Operations Research* 28, pp. 594-602 (1980).

16. W. Whitt, "The Queueing Network Analyzer," *Bell Systems Technical Journal* 62(9), pp. 2779-2815 (1983).

18. S.B. Gershwin, "An Efficient Decomposition Method for the Approximate Evaluation of Tandem Queues with Finite Storage Space and Blocking," *Operations Research* 35(2), pp. 291-305 (1987).

18. M.B.M. De Koster, "Estimation of Line Efficiency by Aggregation," *International Journal of Production Research* 25(4), pp. 615-626 (1987).

19. P.P. Bocharov, "Approximate Method of Design of Open Nonexponential Queueing Networks of Finite Capacity with Losses or Blocking," *Automation and Remote Control (English translation)*, pp. 55-65 (1987).

20. Y.F. Choong and S.B. Gershwin, "A Decomposition Method for the Approximate Evaluation of Capacitated Tranfer Lines with Unreliable Machines and Random Processing Times," *IIE Transactions* 19(2), pp. 150-159 (1987).

21. M. Reiser, "A Queueing Network Analysis of Computer Commumication Networks with Window Flow Control," *IEEE Transactions on Communications* 27(8), pp. 1199-1209 (1979).

22. M. Reiser, and S.S. Lavenberg, "Mean-Value Analysis of Closed Multichain Queueing Networks," *Journal of the Association for Computing Machinery* 27(2), pp. 313-322 (1980).

23. R. Suri, R.R. Hildebrant, "Modeling Flexible Manufacturing Systems Using Mean-Value Analysis," *Journal of Manufacturing Systems* 3(1), pp. 27-38 (1984).

24. G. Cohen, D. Dubois, J.P. Quadrat, and M. Viot, "A Linear System Theoretic View of Discrete Event Processes and its Use for Performance Evaluation in Manufacturing," *IEEE Transactions on Automatic Control* AC-30, pp. 210-220 (1985).

25. J.A.S. Resing, R.E. de Vries, G. Hooghiemstra, M.S. Keane, and G.J. Olsder, "Asymptotic Behavior of Random Discrete Event Systems," *Proceedings of the 28th IEEE Conference on Decision and Control*, Tampa, Florida (1989).

26. K. Barten, "A Queueing Simulator for Determining Optimum Inventory Models in a Sequential Process," *Journal of Industrial Engineering* 13(4), pp. 245-252 (1962).

27. M.C. Freeman, "The Effects of Breakdowns and Interstage Storage on Production Line Capacity," *Journal of Industrial Engineering* 15(4), pp. 194-200 (1964).

28. D.R. Anderson, C.L. Moodie, "Optimal Buffer Storage Capacity in Production Line Systems," *International Journal of Production Research* 7(3), pp. 233-240 (1969).

29. E. Kay, "Buffer Stocks in Automatic Lines," *International Journal of*

Production Research 10(2), pp. 155-165 (1972).
30. J.G. Shanthikumar, and R.G. Sargent, "A Unifying View of Hybrid Simulation/Analytic Models and Modeling," *Operations Research* 31(6), pp. 1030-1052 (1983).
31. Y.C. Ho, M.A. Eyler, and T.T. Chien, "A Gradient Technique for General Buffer Storage Design in a Production Line," *International Journal of Production Research* 27(6), pp. 557-580 (1979).
32. Y.C. Ho, M.A. Eyler, T.T. Chien, "A New Approach to Determine Parameter Sensitivities of Transer Lines," *Management Science* 29(6), pp. 700-714 (1983).
33. Y.C. Ho, and X.R. Cao, "Perturbation Analysis and Optimization of Queueing Networks," *Journal of Optimization Theory and Applications* 40(4), pp. 559-582 (1983).
34. X.R. Cao, "Realization Probability in Closed Jackson Queueing Networks and its Application," *Advances in Applied Probability* 19, pp. 708-738 (1987).
35. Y.C. Ho, X.R. Cao, and C.G. Cassandras, "Infinitesimal and Finite Perturbation Analysis for Queueing Networks," *Automatica* 19(4), (1983).
36. C.G. Cassandras, "On-Line Optimization of a Flow Control Strategy," *IEEE Transactions on Automatic Control* AC-32(11), pp. 1014-1017 (1987).
37. H. D'Angelo, M. Caramanis, S. Finger, A. Mavretic, Y. Phillis, and E. Ramsden, "Event-Driven Model of Unreliable Production Lines with Storage," *International Journal of Production Research* 26(7), pp. 1173-1182 (1988).
38. V.S. Kouikoglou, and Y.A. Phillis, "An Efficient Discrete-Event Model for Production Networks of General Geometry," *Proceedings of the 29th IEEE Conference on Decision and Control*, Honolulu, Hawaii (1990).
39. V.S. Kouikoglou, and Y.A. Phillis, "An Exact Efficient Discrete-Event Model for Production Lines with Buffers," *Proceedings of the 28th IEEE Conference on Decision and Control*, Tampa, Florida (1989).
40. M. Caramanis, "Production System Design: A Discrete Event Dynamic System and Generalized Benders' Decomposition Approach," *Internaitonal Jounral of Production Research*, 25(8), pp. 1223-1234 (1987).
41. V.S. Kouikoglou, and Y.A.Phillis, "An Exact Discrete-Event Model and Control Policies for Production Lines with Buffers," *IEEE Transactions on Automatic Control* AC-36(5), pp. 515-527 (1991).
42. J. A. Buzacott, and L.E. Hanifin, "Models for Automatic Transfer Lines with Inventory Banks: A Review and Comparison," *AIEE Transactions* 10(2), pp. 197-207 (1978).
43. D.R. Smith, "Optimal Repairman Allocation-Asymptotic Results,"

Management Science **24**, pp. 665-674 (1978).

44. K.F.Li, "Serial Production Lines with Unreliable Machines and Limited Repair," *Naval Research Logistics* **34**, pp. 101-108 (1978).

45. X.R. Cao, and Y.C. Ho, "Sensitivity Analysis and Optimization of Throughput in a Production Line with Blocking," IEEE Transactions on Automatic Control **AC-32**(11), pp. 959-967 (1987).

SCHEDULING, OPTIMIZATION AND CONTROL IN AUTOMATED SYSTEMS

S. DAVID WU
Department of Industrial Engineering
Lehigh University
Bethlehem, PA 18015

RICHARD A. WYSK
Department of Industrial Engineering
Texas A&M University
College Station, TX 77843

I. INTRODUCTION

The performance of an automated manufacturing system not only relies on state-of-the-art machinery but also on the effective planning and control of day-to-day operations. Of critical importance is the scheduling of production operations and the control of planned activities. Traditional methods to shop scheduling typically involve the generation of a static, pre-computed schedule using optimization or heuristic methods. The static schedule is generated based on certain managerial or operational objectives subject to shop constraints. In real-world applications, however, the moment a schedule is released to the shop, various uncertainty aspects of the system will render this schedule nothing but a far reaching guideline for shop operations. As a result, planning activities in the shop relies heavily on human intervention. Typically a foreman or a plant manager makes prompt operational decisions based on their best judgment at the moment, and a human scheduler has to be constantly on call in order to adjust, update, or even reproduce the operations schedule. The obvious drawback of such an approach is that it becomes very difficult to evaluate the impact of each operational decision, and constant human interaction is necessary to operate the system. Such a scheduling philosophy, although needs improvement, works in traditional manufacturing. In an automated system, however, constant human intervention is not feasible. A poorly justified decision may cause significant impact such as the jamming or blocking of material flows, the excessive accumulation of work-in-process, or the poor utilization of machines. In a worst case, the marginal benefit brought by automated machinery may be eliminated by inefficient planning.

The scheduling and control in automated systems involves different issues from traditional systems. For instance, timely decisions which take into account realistic constraints are a lot more crucial than, say, optimized makespan. Also, since human intervention will be minimal, it becomes very important to consider the uncertain aspects of the system. Machine disruptions, material shortages, and processing time variations, for example, need to be considered explicitly. Moreover, scheduling should be considered an integral part of the overall planning system. Other elements of the planning system, such as long term strategic planning, inventory control, part loading, material handling, and tooling/fixturing planning need to be taken into account. Interface and integration issues must be considered in order to provide a consistent and coherent treatment to a wide range of planning activities.

For a successful implementation of automated manufacturing, there is an immediate need for all the above issues to be addressed. Unfortunately, the notorious curse of computational complexity makes shop scheduling, along with other combinatorial problems, both difficult to solve and computationally demanding. Consequently when developing solution methodologies, although all the above issues are important, trade-offs have to be made. Techniques have been developed that try to satisfy all the constraints on the shop floor and yet ignore uncertain factors in the future. On the other hand, there are techniques that considers intensively the stochastic behavior of the environment at the expense of restrictive assumptions or mathematical intractability.

In summary, crucial issues involved in the scheduling and optimization of automated systems include the following: 1) a realistic set of constraints must be taken into account, 2) the uncertain nature of the system must be treated, 3) the embedded relationship to other planning activities must be considered and 4) the trade-offs imposed by computational limitations must be addressed.

In the past decade, intensive research has been devoted to study different aspects of the above problems and numerous analytical as well as experimental techniques have been developed. These techniques, although all have their own limitations, provide a crucial scientific basis for future developments. In the next section, an overview of the scheduling, optimization and control issues in automated systems is provided. A brief review of the existing techniques is also given.

II. Critical Issues of Scheduling, Optimization and Control in Automated Systems

The scheduling and control models of automated manufacturing systems are often classified based on their planning horizon and corresponding decision variables (c.f., [1],[2],[3],[4]). In their 1986 review, Kalunte et al. [1] classified the modeling of automated systems into the following four levels:

(1) strategic analysis and economic justification, which provides long-range, plant-wide strategic plans;

(2) facility design, in which strategic business plans are coalesced into a specific facility design to accomplish long-term managerial

objectives;

(3) intermediate range planning, which encompasses decisions related to master production scheduling and deals with a planning horizon from several days to several months in duration;

(4) dynamic operations planning, which is concerned with the dynamic, minute-to-minute operations of the FMSs.

According to Kalunte's classification, this paper addresses the combined scheduling and control problems in the dynamic operations planning level. This level of planning is also known as short-term production scheduling or on-line control and scheduling in a manufacturing control system. The short-term scheduling activities are normally conducted in the cell level of the control hierarchy [4].

A. Constraint Satisfaction

For mathematical and computational tractability, analytical models for scheduling and control are often confined to simplified constraints and assumptions. However, in automated systems, a scheduling and control procedure can be only executed when all the practical constraints are satisfied. Ad hoc dispatching heuristics are frequently used since most constraints can be satisfied without much effort. Nonetheless, when these heuristics are used with a extremely myopic view of the system, the overall system performance will rapidly deteriorate and very little control will be imposed. As a consequence, the effort of planning and control is reduced to a task of "keeping the system going."

An important issue is, therefore, how to treat the constraints one may confront in real situations and at the same time take a global view of the long-term system performance. Numerous researchers have shown that this can be done by taking advantages of discrete optimization techniques which have evolved in the past twenty years. Despite the fact that the constraints treated in optimization models are often limited, these models provide invaluable insights to the overall problem which may be used for the development of efficient approximation or heuristic techniques; in which case, realistic constraints may be taken into consideration. For instance, scheduling techniques have been developed in recent years which use sophisticated models for the analysis of overall system performance. Dynamic dispatching heuristics dictated by the model are then used to sequence jobs.

B. Uncertainty Issues in Scheduling and Control

Schedules for short to medium term operations are typically produced *a priori* in order to direct production operations and to support other planning activities such as tooling, raw material delivery and resource allocation. Unfortunately, as soon as the schedule is released to the shop, myriad disturbances will arise which render the *a priori* schedule obsolete. From the scheduling/control point of view, disturbances of concern may include operational delay, short-term machine break-down, temporary material shortage, etc. These minor system disruptions are normally assumed to occur randomly. In an automated system, it is important to analyze these disturbances such that their impact to the planning activities can be minimized. In a well controlled environment, many of the disturbances are known in advance, or can be predicted in the short-term. For instance, regular machine maintenance may be determined ahead of time and can be included as part of the operations schedule. Temporary material shortages may be predicted accurately in the short-term and can be used to locally adjust the schedule (and thus increase its stability). As for unforeseen random disruptions such as power failure or sudden machine break-downs, rescheduling methodologies may be developed which globally update the *a priori* schedule when a disruption occurs. To maintain system stability, the updated schedule should adhere closely to the original schedule.

Most previous approaches to system disruptions can be divided into two categories: (1) complete rescheduling, and (2) match-up scheduling. The complete rescheduling method applies standard scheduling procedures to produce a new schedule on occurrence of a disruption. The new schedule, however, may deviate significantly from the pre-schedule and causes drastic effects on other system activities that are planned based on the pre-schedule. The match-up scheduling method, on the other hand, computes a transient schedule on occurrence of a disruption and this transient schedule will match up with the pre-schedule in a finite amount of time.

C. Integration Issues

Besides operations scheduling, the planning and control of automated systems often involves complex, interrelated issues such as resource allocation, tool/fixture planning and part loading. In order to maximize performance, a consistent treatment of all these issues is essential. In the past decade, numerous researchers have attempted different aspects of the planning and control problems and a wealth of mathematical as well as experimental frameworks have been established. Because of the inherit difficulty of the problem, most of the analytical models are setup to study a certain tractable subset of the overall problem. This is achieved through some form of aggregation, decomposition, or in some cases, simplification of reality. While the analytical development is extremely important in establishing the foundation for scientific research, direct implementation of these methods in an industrial setup may confront major difficulties. This is due to the fact that each analytical solution examines a limited aspect of the overall problem, integrating these techniques in a consistent, coherent fashion becomes a critical issue. Typically, highly skilled technical personnel are required for the task.

In automated systems, each planning task must be considered as an integral part of the overall planning activities. For instance, when regenerating a schedule upon system disruption, the impact of schedule changes to other planning activities must be carefully analyzed. Such impact ranges from the need for varying the raw material delivery and tooling/fixturing setups, to the possibility of excessive work-in-process and mis-fetched machining part programs. Explicit relationships have to be identified and established among planning tasks such that prompt decisions can be made to update, modify and redirect planned actions.

D. Computational Complexity

Scheduling and control problems are by their basic nature combinatorial. Like other well known NP-hard combinatorial problems such as traveling salesman and knapsack, no efficient polynomial time algorithms exist that solve the problem optimally. Implicit enumeration procedures such as branch and bound or dynamic programming are confined to small-size problems of limited practical importance. In the earlier years of combinatorial research, the computational limitations of computer algorithms made even highly simplified models hard to solve, not to mention models that considered such factors as system uncertainty, model integration and practical constraints. In the past decade, the computing power of personal computers and workstations has advanced in an unimaginable rate. Some computationally demanding algorithms, which would have taken unobtainable computer times in the past, have become feasible. New probabilistic search techniques such as simulated annealing, genetic algorithms and tabu search methods, are examples of those computer algorithms that take full advantage of the technological advancement. An exciting implication is that many large-scale combinatorial optimization problems have become "solvable," or more precisely, good quality solutions can be found in reasonable computer time.

Despite the advances in computing power, the issue of computational complexity remains. For instance, an exponential time algorithm of order, say $O(2^n)$, when the problem size is n=40, a medium size computer can solve it in one day. If a computer ten times faster is used, with the same amount of time one can only solve a problem up to n=43. Clearly polynomial time algorithms take much better advantage of technology.

Given the above reasons, it becomes apparent that in order to solve practical size problems and to take better advantage of the technological advancement, polynomial time heuristics or approximation techniques should be the focus of development. With little surprise, the research community in scheduling and optimization has already headed toward this direction. Numerous heuristic procedures were developed in recent years which gives high quality solutions in reasonable computer time.

III. Existing Techniques in Scheduling and Control

Impressive advancement has been made in the past decade for the solution of scheduling and control problems. Researchers in Operations Research, Production Systems and Manufacturing Systems Engineering have developed methodologies that addresses important aspects of the problem. In the following, a brief overview of these techniques are given. The overview is not intended as a survey of the entire area, instead, techniques that are closely related to the above issues are discussed. These techniques are divided into the following categories : 1) the control theoretic approach, 2) schedule optimization under uncertainty, and 3) simulation-based techniques.

A. The Control Theoretic Approaches

The control and scheduling of a system, in many aspects, is similar to the control of individual devices or subsystems. Over the years, control theorists have attempted to extend various models developed in control theory for the problem of system control. A significant set of conceptual frameworks as well as solution methodologies were developed as a result of this effort.

1. Discrete Event Dynamic Systems

The on-line control activities in an automated system can be modeled as a discrete event dynamic systems (DEDS), where the occurrences of system disturbances are treated as distinct events and one of a finite number of control actions are taken by the controller at each decision point. Markov decision processes (MDPs) are traditionally the analytical approach to these problems. Work in this area has achieved significant success in modeling techniques and solution strategies. The major drawback of MDPs is the restrictive assumptions that must be satisfied for the models to be valid. For example, it is commonly assumed that the set of states and the transition probability matrix (t.p.m.) are available at the time of analysis, and the transition probability must be stationary. These models are generally considered in the areas of "statistical decision analysis" or "adaptive decision making under uncertainty." Decision trees are commonly used to represent problems of this nature. Dreyfus and Law [5], among others, referred to this class of problems as "optimization problems involving learning." They described the use of stochastic Dynamic Programming (DP) and Bayesian decision theory to solve a decision tree. Where "learning" is achieved by using Bayes' law to define the recursive relation in the DP formulation such that an observed event could modify the uncertainty of a future event. Nonetheless, they stated that "... the incorporation of learning into a dynamic-programming model entails an increase in the number of state variables. This usually precludes computational or analytical solution (for realistic models)." [5, p. 195].

A recent development in decision theory, termed "influence diagrams," also attempts the issue of decision making under uncertainty (c.f., [6][7][8][9]). An influence diagram can be described as a carefully structured decision network where conditional dependence of correlated decisions can be modeled and analyzed. Solution algorithms were also developed [7] to solve some of the decision models. However, in an influence diagram, a finite set of outcomes and their associated conditional probability distribution have to be defined explicitly for each node (e.g., a decision variable) in the network. For on-line control problems, some of the states (outcomes of a decision) may be undefined at the beginning, but are determined as the analysis proceeds. In other words, the set of outcomes and their conditional probability distribution is unknown at the beginning of the decision process and depend strongly on the previous decisions and their outcomes. Consequently, utilizing an influence diagram in a control structure needs further investigation yet.

Wu [10] and Leon et al. [11] have suggested a control structure for DEDS using a game-theoretic approach. They modeled the problem of on-line control as a game against nature. The nature makes stochastic moves which represents system disturbances, the controller return moves that will maximize system performance. One advantage of the game-theoretic approach is that the states of the system and their associated probability are not stationary, nor are they predetermined. By solving this game tree, optimal control policies (moves) for the DEDS can be determined by a limited look-ahead of future events.

2. Hierarchical Decomposition Techniques

The integration, computation and some of the uncertainty issues involves in automated systems can be successfully treated by decomposing the overall problem into layers of subproblems. The general principle of hierarchical decomposition is to establish communication among subproblems by passing top-down control input and bottom-up status feedback. Based on the input and the feedback, the subproblem in each layer can be sufficiently solved. This general philosophy of hierarchical decomposition has been adopted by numerous researchers as a solution framework. In 1986, Jones and McLean [4] at the National Bureau of Standard (now National Institute of Standards and Technology) had proposed a five-layer hierarchical control architecture as a standard for the control of the various production and support activities needed to drive an automated factory.

An obvious advantage of hierarchical decomposition is the explicit integration of control activities in the system and the reduced computational complexity. Very often uncertainty issues are also treated within the hierarchy. Hildebrant [12] examined the FMS scheduling and control problem by considering a system with failure-prone machines. He proposed a three-level hierarchy: the top level utilized a nonlinear mathematical program to minimize makespan with the part mix and part routing predetermined for each machine-failure condition. The second and third levels determine loading schedules for the parts to maximize the average production rate and to minimize waiting time, respectively. Buzacott and Yao [13] indicated that since Hildebrant ignored the transient period following each machine-failure, his control strategies may not be optimal in a real situation.

Similar work by Kimemia [14] and Kimemia and Gershwin [15] proposed a three-level control hierarchy which involves: 1) a Flow Controller which calculates/regulates short-term production rates for each part type; 2) a Routing Controller which determines the part routing in order to meet the production rate dictated by the Flow Controller; and 3) a Sequence Controller which dispatch parts into the system to maintain the flow rates and routing chosen by the previous controllers. It differs from Hildebrant's [12] scheme in that "... flow rate decisions are made on the basis of the current inventory levels as well as the current set of working machines" [15, p. 355], while in Hildebrant's model, the inventory levels are not considered.

Tsitsikilis [16] examined dynamic routing policies to

compensate for machine breakdown using an approach that modeled queues as levels of continuous state-variables.

Odrey and Wilson [17] developed a cell and a shop level control structure for FMSs. The structure is composed of planning and control methodologies, the scheduling problem being represented by a state-space formalisms. The planning module utilizes steady-state periodic loading of part types to minimize cycle time. The module also determines the spectral properties of the system (e.g., the transient-state duration, periodicity, etc.) and adjusts these properties. Finally the steady-state limit times for all activities during the planning horizon are calculated. The steady-state periodic loading scheme is followed until a system disruption occurs. The control module then generates an interim scheduling policy to prevent the accumulation of in-process-inventory until the down machine becomes operational, at which time the system enters the transient phase. In the transient phase, the control module enforces a set of operational strategies to bring the system back to the steady-state periodic schedule as quickly as possible.

B. Scheduling and Control Under Uncertainty

Production scheduling problems have provided an area of extensive research in the fields of Combinatorial Optimization for years. Work in this area has provided a wealth of information on solution strategies and approximation algorithms for determining optimal or near optimal schedules. As discussed earlier, most of the literature in this field deals with static scheduling models, that is, generating a deterministic schedule over a considerable time frame assuming that all problem characteristics are known. Static models typically encounter difficulty in practice due to the dynamic nature of the shop floor caused by inevitable random disruptions. Machine break-down, new job arrival, etc., may dislocate the scheduled jobs from the prescribed sequence or start times and render the predetermined schedule obsolete. Buzacott and Yao [13], Suri and Whitney [2], and Graves [18] have all stressed the importance of studying such transient system behavior for the operation and control of automated systems.

As discussed in previous section, the control of a production schedule can be considered as a sequential decision process under uncertainty, commonly modeled by discrete-state Markov decision processes (MDPs). Stochastic dynamic programming has been the traditional approach to MDPs [19][20]. Nevertheless, as indicated by CONDOR [21]: "Realistic MDP models of real-world decision making ... are often computationally demanding, if not intractable, due to the inherent curse of dimensionality associated with dynamic programing." [21, p.629-30] They further suggested: "Application of heuristic search algorithms (such as A∗, AO∗, and several extensions) to MDPs and other models of sequential decision making under uncertainty is an important research direction." [21, p.630]

1. Non-Myopic Job Dispatching Policies

The uncertainty factors typically considered for scheduling and control include random job arrival, processing time variations, and system disturbances such as machine breakdown, material shortage, power failure, etc. As discussed earlier in Section II, ad hoc dispatching heuristics are often used in practice such that practical constraints can be satisfied and the uncertainty issues of the system can be avoided. However, to ensure control of the system, it is desirable to have dispatching policies which are dynamic by nature and at the same time maintaining a non-myopic view of the long - term system performance.

Work in stochastic scheduling has dealt with scheduling models that explicitly reflect the uncertain nature of the available information (e.g., job processing time may vary), and thus generate optimal dispatching policies. Research such as Pinedo and Ross [22],

Glazebrook [23][24][25][26], and Pinedo [27] fall in this category. Glazebrook examined the effects of machine breakdowns in a single-machine problems using a cost-discounted Markov process [24], and a Semi-Markov model [25]. However, to date, the existing techniques are for relatively small problems.

Roundy et al. [28] proposed a two-module scheduling system: a scheduling module determines the cost of using a machine at a given time, and a dispatching module dispatches the jobs based on the established cost information. A similar approach proposed by Morton et al. [29] and Morton et al. [30] suggest an evolving cost based method where the job dispatching decision is made based on the pre-computed machine usage cost, tardiness cost, etc. The work cited above provides an excellent hybrid of static scheduling and dynamic dispatching. The primary advantage of the approach is that the off-line scheduling module maintains a global view of the system such that dynamic dispatching policies are always restricted by the pre-optimized schedule. Although they do not consider disturbances explicitly in the model, the cost (or the schedule) is recomputed when a major disruption takes place.

2. Rescheduling and Match-Up Scheduling Methods

One line of research has been to study the implementation of a pre-computed off-line schedule subject to various system disturbances. A premise of this set of studies is that a schedule must be pre-computed to optimize system performance, and to serve as a basis of other planning activities.

Work by Bean and Birge [31], Birge and Dempster [32], Gallego [33][34] have investigated "match-up" scheduling methods that compute a transient schedule after a machine disruption. This transient schedule will eventually match up with the pre-schedule in a finite amount of time. They stressed that this approach has intuitive appeal since material flow in the system are planned based on the pre-schedule. However, as they indicated, the imposition of a finite match-up time may cause significant delay in the transient period, and bounds were developed for the delay. A difficult aspect of the method is to determine the match-up time.

On the other hand, Yamamoto and Nof [35] investigated the efficiency of implementing a scheduling/rescheduling procedure in a real-time, computer-controlled environment. The procedure rescheduled the system on each significant operational change (e.g., machine break-down). They reported advantages of between 2.5% to 7% when compared to a fixed sequencing and a priority dispatching procedure, respectively. A drawback of the method is that rescheduling changes the pre-schedule significantly which may incur major cost impact to other planning activities.

Wu [10], Wu and Leon [36] and Leon et al. [11] have suggested a three-stage decision structure for scheduling and control. In the first stage, an off-line schedule is generated to optimize long-term system performance considering the presence of system disruptions. Since the off-line schedule serves as a basis for other planning functions, it is desirable to control the execution of this schedule in real-time. In the second stage, an on-line control structure performs limited lookahead of potential disturbances in the future and control policies are prescribed which adjust the schedule according to this information. The overall objective is to utilize the on-line information and resolving a contingency plan which minimizes the damage caused by system disturbances. The on-line information is provided as a time-varying function of machine breakdown distributions. In the third stage, where major disturbances render on-line control impractical, a rescheduling procedure is used to update the off-line schedule. The new schedule is designed to maximize long-term performance and at the same time adheres closely to the original schedule. This is achieved by solving a bicriterion scheduling problem with the original objective and an additional objective of minimizing deviation from the original schedule [37].

3. Robust Scheduling

One line of research has been to study the "robustness" of schedules by evaluating the expected performance of a schedule on occurrence of future disruptions. Leon et al. [38] have developed robustness measures for job shop scheduling with makespan as objective. A delay function is developed which relate the slack times exist in the schedule with the arrival times and durations of machine disruptions. Expected delay and expected makespan based on the arrival times and duration distribution is then computed as the robustness measure. They further apply the measure to generate robust schedules using a local search algorithm.

Another line of research addresses the issue of scheduling robustness by characterizing the planning horizons for deterministic scheduling models, where a planning horizon establishes the insensitivity of the current scheduling decisions to future information beyond the horizon [18]. Morton [39], for example, has found conditions for determining planning horizons for a class of convex-cost scheduling problems. Earlier work by Wagner and Whitin [40], and Zabel [41] also provide some important results for lot-sizing problems.

Related work by Baker [42], and Baker and Peterson [43] examined the efficiency of obtaining scheduling decisions from optimizing a finite, multi-period model and implementing those decisions on a rolling basis (i.e., only a model decision for the immediate planning period is implemented; in the next period, the schedule is re-optimized based on revised and additional shop information, and again only the immediate period's decisions are used). They found that under reasonable conditions such rolling schedules are quite robust.

Graves [18] indicated that as a consequence of implementing the rolling schedules, the schedules for future periods are repeatedly changed. This instability can be costly if the schedule is used as a basis for other planning systems (e.g., inventory control systems). Some preliminary work attempting to incorporate these costs into the scheduling decision is reported by Carlson et al. [44]. They proposed a modification to a MRP-based dynamic lot-sizing system to reflect the cost of a schedule change.

C. Simulation Based Methods

In recent years, discrete event simulation has been proposed by many as a practical approach to the control and scheduling problems. The basic idea is to use simulation for "limited lookahead" such that what-if analysis can be performed to evaluate future alternatives subject to available status information. Along the same line, Shanthikumar and Sargent [45] have suggested various hybrid simulation/analytic models. One class of hybrids they suggested use a simulation model as an overall model and some or all of its parameters are computed from the solution procedure of an analytic model. A primary advantage of the approach is that simulation models are able to accommodate deterministic as well as stochastic information available from the shop. More importantly, simulation may serve as a means of integrating the pieces of analytical models each developed for a limited aspect of the problem. For example, part loading algorithms, material handling policies and operations scheduling procedures may be integrated under a common simulation model. The overall system performance can be assessed accurately and parameters can be passed among models for further tuning and improvement. The main obstacles of using such simulation structure in an industrial setup are as follows: 1) development of the simulation model often requires considerable effort, and 2) modifying or updating the simulation model for system changes are costly. To resolve the above problems, simulation "generators," which translate system models into programming codes, are proposed by many researchers.

Simulation based analysis and control is not a replacement for the analytical methods one may find from the literature. Rather, simulation and simulation generators provides a practical means of integrating the pieces of analytical development. It may not be possible for such integration to be "globally optimized," and some mathematical elegance will sure be sacrificed. Nonetheless, a consistent heuristic treatment to the overall problem will be provided and thus solutions of practical significance will result.

In a 1984 survey, Mathewson [46] summarized the earlier work of program generators and their extensions to discrete event simulation. Based on his definition, a program generator is "a tool to aid in the production of computer-coded representation of a logical model [46, p.3]." In the case of discrete event simulation, the code will be generated based on some system model represented by networks, event diagrams, entity cycle diagrams, or other modeling tools. Majority of the work in simulation generators transfer the system model into specific simulation languages. The DRAFT family of generators [47], for instance, generates SIMON/FORTRAN, GASP II, SIMSCRIPT II.5, SIMULA, and 2900 ACSL codes based on entity cycle diagrams. Haddock [48] described a special purpose generator which generates SIMAN code based on a specific FMS model.

Another line of research addresses the issues of using simulation as a means of on-line control and decision making. ElMaraghy and Ho [49] summarized the potential use of such approach as follows: 1) to test specific system control strategies, 2) to identify the critical control elements, and
3) to offer support in rescheduling the released orders or changing the part-mix. Although the general concept has been broadly suggested, two major problems have delayed this technique from fully implementation: 1) The generation/updating of simulation models is time-consuming and costly, 2) system data required to build a model must be kept current.

Wu and Wysk [50][51] proposed a control structure using expert systems and simulation. The expert system keeps tract of system status, proposes scheduling alternatives based on the status, generates corresponding SIMAN simulation codes, then evaluates the alternatives using the simulation model. Other researchers have also tried to create simulation environments which integrate knowledge-based software with databases, statistics software, and computer graphics. Reilly et al. [52], for instance, developed a integrated system architecture which includes a LISP-based model builder, a SLAM-based model executor, a database record keeper, and a statistical package (i.e. SAS) for output analysis. Ben-Arieh [53] described a knowledge-based simulation environment using PROLOG. The environment includes a database which maintains static and dynamic system information, a knowledge base which offers the user a programming library to organize the simulation model, and a simulator which execute the model.

Another important used of simulation is in the analysis of discrete event dynamic systems. Different from the set of research cited above, this set of research uses simulation as an experiment tool for detailed system analysis. Since stochastic phenomena are considered, simulation experiments often involve a large Monte Carlo run on the computer which is computationally difficult for real-time applications. An approach called "perturbation analysis" [54] was proposed to analyze DEDS by using simulation intelligently. The basic idea is to performs a "most meaningful" analysis of future events subject to limited computing power. The basic technique they used is to study the sensitivity of some system performance to a particular set of parameters using simulation or direct observation of the actual system. Based on perturbation analysis, Suri and Cao [55] and Suri and Dille [56] study the design problems in FMSs through analyzing some specific system parameters. Other applications of perturbation analysis can be found in [57][58][59].

IV. Conclusions

In this paper, the scheduling, optimization and control issues of automated systems are briefly addressed. Some existing research that addresses these issues is reviewed. From the review, one may agree that the research for the planning and control of automated systems represents a subset of the production planning/control literature that provide treatment to issues that are unique to automated system. Namely, the issues of satisfying practical constraints, coordinating different planning functions, recovering from uncertain events and providing computationally feasible solution algorithms. These issues may seem insignificant for tradition systems, they are extremely critical to automated environment since constant human intervention can be no long assumed and relied on.

References

1. Kalunte, M.V., Sarin, S.C. and Wilhelm, W.E., "Flexible Manufacturing Systems: A Review of Modeling Approaches for Design, Justification and Operation," Flexible Manufacturing Systems: Methods and Studies, A. Kusiak (Editor), Elsevier Science Publishers B.V. (North-Holland), 1986.

2. Suri, R. and Whitney, C.K., "Decision Support Requirements in Flexible Manufacturing," SME Journal of Manufacturing Systems 3, 61-69 (1984).

3. Simpson, J. A., Hocken, R. J. and Albus, J. S., "The Automated Manufacturing Research Facility of the National Bureau of Standards," Journal of Manufacturing Systems 1, 17-32 (1982).

4. Jones, A. T. and Mclean, C.R., "A Proposed Hierarchical Control Model for Automated Manufacturing Systems," Journal of Manufacturing Systems 1, 15-25 (1986).

5. Dreyfus, S.E. and Law, A.M., The Art and Theory of Dynamic Programming, Academic Press, 1977.

6. Schachter, R.D., "Evaluating Influence Diagrams," Operations Research 34, 871-882 (1986).

7. Schachter, R.D., "Probabilistic Inference and Influence Diagrams," Operations Research 36, 589-604 (1988).

8. Howard, R.A. and Matheson, J.E., "Influence Diagrams, 1981." In the Principles and Applications of Decision Analysis, Vol. II, Howard, R.A. and Matheson, J.E. (eds.). Strategic Decision Croup, Melo Park, CA, 1984.

9. Howard, R.A., "Decision Analysis: Practice and Promise," Management Sci. 34, 679-695 (1988).

10. Wu, S. D., "Scheduling, Control and Rescheduling Methodologies for Uncertain Manufacturing Environments," in Proceedings of the 1991 NSF Design and Manufacturing Systems Conference, Austin Texas, 1991.

11. Leon, V. J., Wu, S. D. and Storer, R. H., "A Game Theoretic Control Approach for Job-Shops in the Presence of Disruptions," Working Paper, Department of Industrial Engineering, Lehigh University, Bethlehem, PA, 1991.

12. Hildebrant, R.R., "Scheduling of Flexible Machining Systems When Machines Are Prone to Failure," Ph.D. Dissertation, Massachusetts Institute of Technology, 1980.

13. Buzacott, J. A. and Yao, D.D., "Flexible Manufacturing Systems: A Review of Analytical Models," Management Science 32, 890-905 (1986).

14. Kimemia, J.G, "Hierarchical Control of Production in Flexible Manufacturing Systems," Ph.D. Dissertation, Report No. LIDS-TH-1215, Laboratory for Information and Decision systems, Massachusetts Institute of Technology, 1982.

15. Kimemia, J.G. and Gershwin, S.B., "An Algorithm for the Computer Control of a Flexible Manufacturing System," IIE Transactions 15, 353-362 (1983).

16. Tsitsiklis, J.N., "Optimal Dynamic Routing in an Unreliable Manufacturing System," LIDS-TH-1069, Laboratory for Information and Decision Systems, Massachusetts Institute of Technology, 1981.

17. Odrey, N.G. and Wilson, G.R., "An On-Line Control Structure for Flexible Manufacturing Systems," Proc. 14th Conference of the NSF Production Research and Technology Program, Ann Arbor, MI, 1987.

18. Graves, C. S., "A Review of Production Scheduling," Operations Research 29, 646-675 (1981).

19. Howard, R.A., Dynamic Programming and Markov Processes, Technology Press and Wiley, 1960.

20. Bertsekas, D., Dynamic Programming: Deterministic and Stochastic Models, Prentice-Hall, Inc., Englewood Cliffs, 1987.

21. Committee On the Next Decade in Operations Research (CONDOR), "Operations Research: The next Decade," Operations Research 36, (1988).

22. Pinedo, M. and Ross, S., "Scheduling Jobs Subject to Non-Homogeneous Poisson Shocks," Management Science 26, 1250-1257 (1980).

23. Glazebrook, K.D., "On Non-Preemptive Strategies in Stochastic Scheduling," Naval Research Logistics Quarterly, Vol. 28, 1981, p. 289-300.

24. Glazebrook, K.D., "Scheduling Stochastic Jobs in a Single Machine

Subject to Breakdowns," Naval Research Logistics Quarterly 31, 1984.

25. Glazebrook, K.D., "Semi-Markov Models for Single-Machine Stochastic Scheduling Problems," Int. J. Sys. Sci. 16, 573-587 (1985).

26. Glazebrook, K.D., "Evaluating the Effects of Machine Breakdowns in Stochastic Scheduling Problems," Naval Research Logistics 34, (1987).

27. Pinedo, M., "Stochastic Scheduling with Release Dates and Due Dates," Operations Research 31, 559-572 (1983).

28. Roundy , R., Herer, Y. and Tayur, S. "Price-Directed Scheduling of Production Operations in Real Time," Proc. 15th Conference of the NSF Production Research and Technology Program, Berkeley, CA, 1989.

29. Morton, T.E., Lawrence, S., Rajagopalan, S. and Kekre, S., "MRP-STAR PATRIARCH's Planning Module," Working Paper, Graduate School of Industrial Administration, Carnegie Mellon University, Pittsburgh, PA, December, 1986.

30. Morton, T.E., Lawrence, S., Rajagopalan, S. and Kekre, S., "SCHED-STAR A Price-Based Shop Scheduling Module," Working Paper, Graduate School of Industrial Administration, Carnegie Mellon University, Pittsburgh, PA, February, 1988.

31. Bean, J. and Birge, J., "Match-up Real-Time Scheduling," Proceeding of a Symposium on Real Time Optimization in Automated Manufacturing Facilities, NBS Publication 724, National Bureau of Standards, 197-212 (1986).

32. Birge, J. and Dempster, M., "Optimality Conditions for Match-Up Strategies in Stochastic Scheduling and Related Dynamic Stochastic Optimization Problems," Working Paper, Department of Industrial Engineering and Operations Engineering, The University of Michigan, Ann Arbor, MI, June 1987.

33. Gallego, G., "Linear Control Policies for Scheduling a Single Facility After An Initial Disruption," Technical Report No. 770, School of Operations Research and Industrial Engineering, Cornell University, Ithaca, NY, January 1988.

34. Gallego, G., "Produce-Up-To Policies for Scheduling a Single Facility After An Initial Disruption," Technical Report No. 771, School of Operations Research and Industrial Engineering, Cornell University, Ithaca, NY, January 1988.

35. Yamamoto, M. and Nof, S.Y., "Scheduling/Rescheduling in the Manufacturing Operating System Environment," IJPR, Vol. 23, No. 4, 1985, p. 705-722.

36. Wu, S.D. and Leon, V. J., "A Decision Structure Using Generalized AND/OR Trees Containing Chance Nodes," Working Paper, Department of Industrial Engineering, Lehigh University, 1990.

37. Leon, V.J. and Wu, S.D., "On Scheduling with Ready-Times, Due-Dates and Vacations," Working Paper, Department of Industrial Engineering, Lehigh University, 1990.

38. Leon, V.J., Wu, S.D. and Storer, R.H., "Robust Measures and Robust Scheduling for Job Shops," Working Paper, Department of Industrial Engineering, Lehigh University, 1990.

39. Morton, T.E., "Universal Planning Horizons for Generalized Convex Production Scheduling," Operations Research 26, 1046-1058 (1978).

40. Wagner, H.M. and Whitin, T., "Dynamic Version of the Economic Lot Size Model," Mgmt. Sci. 5, 89-96 (1958).

41. Zabel, E., "Some Generalizations of an Inventory Planning Horizon Theorem," Mgmt. Sci. 10, 465-471 (1964).

42. Baker, K.R., "An Experimental Study of the Effectiveness of Rolling Schedules in Production Planning," Decision Sci. 8, 19-27 (1977).

43. Baker, K.R. and Peterson , D.W., "An Analytic Framework for Evaluation Rolling Schedules," Mgmt. Sci. 25, 341-351 (1979).

44. Carlson, R.C., Jucker, J.V. and Kropp, D.H., "Less Nervous MRP Systems: A Dynamic Economic Lot-Sizing Approach," Mgmt. Sci. 25, 754-761 (1979).

45. Shanthikumar, J.G. and Sargent, R.G., "A Unifying View of Hybrid Simulation/Analytic Models and Modeling," Operations Research 31, 1030-53 (1983).

46. Mathewson, S.C., "The Application of Program Generator Software and Its Extensions to Discrete Event Simulation Modeling," IIE Transactions 16, 3-18 (1984).

47. Mathewson, S.C., "A DRAFT II?SIMON Manual," Proceedings of

the Department of Management Science, Imperial College, London 1982.

48. Haddock, J., A Simulation Generator for Flexible Manufacturing Systems Design and Control, IIE Transactions 20, 22-31 (1988).

49. ElMaraghy, H.A. and Ho, N.C., "A Simulator for Flexible Manufacturing Systems with Graphical Animation," Computers in Engineering Vol. 1, Proceedings of Second International Computer Engineering Conference in San Diego, ASME, NY 1982.

50. Wu, S. D. and Wysk, R. A., "An Application of Discrete-Event Simulation to On-Line Control and Scheduling in Flexible Manufacturing," IJPR 27, 1603-1623 (1989).

51. Wu, S. D. and Wysk, R. A., "An Inference Structure for the Control and Scheduling of Manufacturing Systems," Computers and Industrial Engineering 18, 247-262 (1990).

52. Reilly, K. D., Jones, W. T., Dey, P., The Simulation Environment Concept Artificial Intelligence Perspectives, Artificial Intelligence and Simulation, 1985.

53. Ben-Arieh D., A Knowledge Based System for Simulation and Control of a CIM, Report, AT&T Bell Laboratories, 1986.

54. Ho, Y.C. and Cao, X.R., "Perturbation Analysis of Discrete Event Dynamic Systems," J. Optim. Theory Appl. 40, 559-582 (1983).

55. Suri, R. and Cao, X.R., " The Phantom Customer and Marked Customer Methods for Optimization of Closed Queueing Networks with Blocking and General Service Times," ACM Performance Evaluation Rev., 243-256 (1983).

56. Suri, R. and Dille, J.W., "On-Line Optimization of FMS Using Perturbation Analysis," Proc., 1st ORSA/TIMS Conf. FMS, Ann Arbor, Michigan, 1984.

57. Ho, Y.C. , Eyler, M.A. and Chien, T.T., "A Gradient Technique for General Buffer Storage Design in a Production Line," IJPR 17, 557-580 (1979).

58. Ho, Y.C. , Eyler, M.A. and Chien, T.T., "A New Approach to Determine Parameter Sensitivities of Transfer Lines," Management Science 29, 700-714 (1983).

59. Ho, Y.C., Suri, R., Cao, X.R., Diehl, G.W., Dille, J.W. and

Zazanis, M., "Optimization of Large Multiclass (Non-Product Form) Queueing Network Using Perturbation Analysis," J. Large Scale Systems 7 (1985).

Performability of Automated Manufacturing Systems

N. Viswanadham   Y. Narahari   R. Ram

Department of Computer Science and Automation
Indian Institute of Science
Bangalore 560 012, INDIA.

Abstract

Automated manufacturing systems (AMSs) can be regarded as *flexible, degradable, fault-tolerant systems*. To evaluate such systems, we need combined measures of performance and reliability, called *performability* measures. The most important performability measures in the AMS context are related to *throughput* and *manufacturing lead time* (MLT) since high productivity and low lead times are prime features determining the competitiveness of AMSs. Performability has a strong relation to the notion of flexibility in manufacturing systems.

Performability modeling is an active research topic in the area of fault-tolerant computing systems. Performability studies are of great interest in the AMS context also, since performability enables to quantify flexibility and competitiveness of a manufacturing system. In this article, we use the conceptualization available in the area of fault-tolerant computing, to formulate performability notions for AMSs. We consider steady-state performability, interval performability, and distribution of performability with respect to throughput and MLT measures. Through several illustrative examples and numerical results, we bring out the importance of performability modeling and evaluation in AMS design. The AMS examples discussed are: (i) a versatile machine center with machine failures and repairs, and (ii) a generic flexible manufacturing system with centralized material handling, where the performance could be described by a closed central server model.

Notation

$[0,t]$	Observation period
u	Arbitrary time instant
$R(u)$	Reliability over $[0,u]$
$A(u)$	Instantaneous availability at u
A	Steady-state availability
S	State space of the structure state process
$\{Z(u) : u \geq 0\}$	Structure state process
$Z(u)$	Structure state at u
i,j	Typical structure states
$p_{i,j}$	Transition probabilities
f_i	Reward rate in structure state i
τ_i	Total time during $[0,t]$ the structure state process stays in state i
s	Initial structure state
$Y_t(s)$	Performability over $[0,t]$ with s as initial state
$Q(t)$	Cumulative production in $[0,t]$
$W(t)$	MLT-related performability in $[0,t]$
x	A desired lower bound on throughput
w	A desired upper bound on MLT
L	Steady-state mean number of parts in the system
W	Steady-state average MLT
λ	Arrival rate of parts
μ	Service rate of parts
α	Failure rate
β	Repair rate

1   Introduction

Automated Manufacturing Systems (AMSs) are highly capital intensive and have short process life cycles. These systems have built-in redundancy in terms of *flexibility* so that the system can effectively cope with demand, design, and product mix changes as well as equipment failures. Guaranteed levels of productivity and payback ratio are essential for survival of such systems in competitive markets. This requires that the system flexibilty is managed to achieve low lead times and high throughput. In this article, we introduce a combined performance and reliability measure for assessing the productivity either in terms of throughput or lead time over a given time horizon.

One factor that has a major influence on system performance is the unscheduled down-time of the equipment due to failures. Traditionally, downtime management is done via reliability, availability, and maintainability theories. Also, performance and reliability issues have been dealt with separately. Reliability and availability are usually computed using Markov and combinatorial models. Discrete event simulation, Markov chains, stochastic Petri nets and queueing networks are popular modeling tools for performance evaluation of the failure-free model. However, AMSs have a high degree of fault tolerance induced by flexibility and can exist in various structure states (modes of operation) during the time intervals of interest and the performance of the system degrades or enhances as the structure of the system changes with failure and repair. Thus, a combined study of performance and reliability using *performability modeling* is highly relevant.

Performability modeling is an important topic of research interest in the area of fault-tolerant computing systems. In this article, we use the conceptualization available in the area of fault-tolerant computing, to formulate performability notions for AMSs. Also through several illustrative examples and numerical results, we bring out the importance of performability modeling and evaluation in AMS design. Studies relating to performability are of immense practical interest in automated manufacturing systems where a variety of part types is to be manufactured in a finite time horizon of a shift period and these systems are failure prone.

A. Automated Manufacturing Systems

A typical AMS comprises a computer controlled configuration of Numerically Controlled (NC) machine tools, assembly stations, and a Material Handling System (MHS) designed to simultaneously manufacture a wide variety of low to medium volumes of high quality products at a low cost. A hierarchical computer control system coordinates the actions of the machines and the MHS and controls the movement of workpieces. Buffer storage at the turn-table of the machine center and also a centralized pallet pool are available to smoothen the flow of the workpieces and also to avoid blocking. Load/Unload stations and inspection stations, automated workpiece storage and retrieval systems, centralized tool stores, and local tool magazines also form subsystems of the AMS.

A raw workpiece is loaded at the Load/Unload station and is routed through the system following the *routing table*. The routing table specifies for each part type the operations to be performed and their precedence relationship, choice of machines for each operation, and time and tools required for each operation. Because the machines are highly versatile, it is possible an operation could be performed on a number of machines and a particular part type could be manufactured via several routes. This leads to the *routing flexibility* which provides resilience against machine breakdowns.

Failures could be hard or soft failures. Some failures such as tool failures, transient controller faults, are on-line recoverable, others such as those involving track, conveyor, automated guided vehicle, and robot may require off-line repair. Hard failures make the system inoperable whereas soft failures such as tool wear create quality control problems. Statistical Quality Control and Statistical Process Control techniques help in diagnosing soft failures.

An AMS is a highly competitive manufacturing strategy and is supposed to adapt quickly to external changes such as design, demand, and product mix changes and also to internal changes such as subsystem failures, processing time variabilities, product quality variations, etc. Dynamic manufacturing requires cutting down waste i.e., all non-value adding operations such as material handling, storage, inspection, rework, and machine breakdowns. Further, competitive and timely delivery are possible only if the waste is cut down and lead times are minimal.

B. Performability Modeling of AMSs

While high quality and low cost are primary attributes of the product, capacity, flexibility, and reliability are AMS layout attributes. Manufacturing Lead Time (MLT) and Throughput are performance measures that depend on system operation. Two other related performance indicators of efficient system operation are Work-In-Process (WIP) and machine utilization. In this paper, we are concerned with *performability*, which is the probability that the system operates at guaranteed levels of performance, in the face of system failures, in terms of a performance measure such as MLT or throughput.

Flexibility is an important attribute of computer controlled manufacturing systems. Flexibility is the ability to respond to change: both internal and external changes defined above. Flexibility is fundamental to achieve competitiveness and to handle risk associated with uncertain markets. Manufacturing systems can have varying degrees of flexibility depending on the versatility of the equipment and the way the equipment is managed. A transfer line is a very effective means for high volume production of a single product. Such lines are highly optimized for a specific product and will not tolerate changes in product design. Failure of a single machine brings the entire system down. On the other hand, a system of identical machines is very flexible and can tolerate internal as well as external changes. More specifically, failure of a machine would not bring the system down, but allows operation in a degraded mode. Two situations can arise with regard to repair in the manufacturing systems. In an unmanned night shift operation, repair is not possible. We are interested in non-repairable system performance in the face of subsystem failures. Repair, manual or automatic, is possible during the day shift period, thus the study of repairable systems is also of interest.

Most AMSs have certain degrees of fault-tolerance mainly because of their flexibility. Thus an AMS is a repairable, degradable, fault-tolerant system. A variety of faults can occur in its subsystems and these would affect the performance of the system. In this paper, we are concerned with determination of the performance of some typical manufacturing system configurations under failure and repair conditions.

Literature on performability modeling of fault-tolerant computer systems is abundant. Beaudry [1] introduced performance related reliability for gracefully degrading fault-tolerant systems. Meyer ([2], [3]) used the term *performability* for the first time and pioneered early research work on this topic. Reibman [4] has given a tutorial introduction to performability modeling in computer systems. Goyal *et al* [5] have given an overview of availability modeling in fault-tolerant computer systems.

It is now common in performability literature to have (i) a *structure state model* or *reliability model* to describe the system evolution as influenced only by failures and repair, and (ii) a *performance model* that would describe the system performance (throughput, response time, etc.) in each individual (structure) state of the structure state model. Performability evaluation will then involve combining the structure state model and the performance models using some computational procedures. Donatiello and Iyer [6] have presented efficient computational algorithms to compute the cumulative distribution function of performability when the structure state model is a general acyclic Markov chain. These results are applicable only to non-repairable systems. More recently, several authors have devised techniques to compute distribution or moments of performability when the structure state model is an irreducible Markov chain, thus taking care of repairable systems. These include Nicola *et al* [7], Iyer *et al* [8], Smith *et al* [9]), and, Silva and Gail [10]. Ciardo *et al* [11] have now extended performability analysis to semi-Markov models.

Reliability and performability analysis of AMSs has not received much attention so far. Kanth and Viswanadham ([12]) have looked at *part* reliabilities and *system* reliabilities for flexible manufacturing systems and described computational methods for obtaining these measures from the routing table for part types. However, their study is confined to reliability studies. Albino *et al* [13] have considered performability notions for AMSs. Their study also describes the two types of models, reliability model and performance model, and gives a computational method for obtaining steady-state throughput in the presence of failures and repairs. They do not consider the computation of the distribution of performability.

C.   Outline of the Article

In Section 2 of this article, we introduce generic performance measures and generic performance modeling tools for AMSs. We then introduce reliability and availability measures for AMSs. Next, we formulate the notion of performability of AMSs using the concepts of structure state process and accumulated rewards. An example of a system with two non-homogeneous, unreliable machines illustrates the definitions in this section.

In Section 3, we discuss the performability modeling of a machine center with a single machine under two cases: infinite buffer and finite buffer. In the former case, steady-state performability is computed using the analysis of an $M/M/1$ queue with breakdowns. In the latter case, the total average throughput in a given interval $[0, t]$ is computed assuming that the machine is not repaired on failure. This would give an interval performability measure.

The well known central server model of Flexible Manufacturing Systems (FMSs) with an AGV and several machines is the topic of discussion in Section 4. Here also, we consider two cases: non-repairable case and repairable case. In the first case, we show the computation of the distribution of throughput and MLT-related performance measures. In the second case, we show how the expected values of performability can be computed.

2   Measures of Performance and Performability

A.   Performance Measures

The performance and competitiveness of an AMS can be captured by certain generic performance measures. Viswanadham and Narahari [14] have listed the following generic measures:

1. Manufacturing Lead Time

2. Work-in-Process

3. Machine Utilization

4. Throughput

5. Capacity

6. Quality

7. Flexibility

8. Performability

The *Manufacturing Lead Time* (MLT) of a product is the total time required to process the product through the manufacturing plant. Ideally, MLT should be equal to the actual machining time. This is only possible with zero inventories, zero material handling, zero breakdowns, and a batch size of unity. These are however ideal conditions and in actual practice, one would wish to minimize the MLT. MLT can be broken down into four components, namely setup time, machine run time, material handling time, and queue time. MLT is an important and an all-encompassing measure of system performance.

The *Work-In-Process* (WIP) is the amount of semi-finished product currently resident on the factory floor. WIP represents an investment by the firm and the inventory costs incurred by many companies are substantial. A large WIP is often the result of poor design, poor forecasting, and poor operation of the factory. A large WIP also leads to large MLTs. Therefore, WIP must be low.

Machine utilization is the fraction of time a machine is producing useful work. High machine utilization is desirable because it amortizes the cost of machinery faster. However, high utilization may also lead to a high WIP. An optimal machine utilization is that which enables to manufacture exactly the right quantity of exactly right products at the right time.

Throughput measures the production efficiency of a manufacturing plant. Throughput depends almost directly on the utilization of the machines and material handling equipment. MLT and throughput constitute an important subset of performance measures that will be discussed in this paper.

The *capacity* of a manufacturing plant is the maximum possible output of the manufacturing process the plant is able to produce over some specified duration. Capacity can be measured as machine hours available per week or the number of units produced per week.

Quality is an important attribute that products are required to possess in order to enhance the competitiveness of a firm. The thrust of new generation manufacturing is towards quality and productivity.

Flexibility is a vital, yet unquantifiable aspect of a manufacturing system. A flexible system is one that is able to respond effectively to change.

Changes could be internal or external. Internal changes include breakdown of equipment, variability of processing times, worker absenteeism, and quality problems. External changes are typically changes in the design, demand, and product mix. The ability to cope with internal changes requires a degree of redundancy in the system whereas the ability to cope with external changes requires that the system be versatile and be capable of producing a wide variety of part types with minimal changeover times and costs to switch from one product to another.

Performability is a composite measure that combines the performance and reliability of a system. Performability is a generic term to describe any performance measure, taking into account failures and repairs of system components such as machines, material handling equipment, and tools. The notion of performability has close relation to that of flexibility.

B.   Performance Models

An AMS can be described as a discrete event dynamical system [15] because changes in the system state are caused by the occurrence of events, at discrete instants of time instead of continuously. Therefore, the evolution of states in an AMS is not described by partial or ordinary differential equations, as in the case of continuous variable dynamic systems. Some examples of discrete events in an AMS include: entry/exit of a part, beginning/completion of processing on a machine, breakdown of a machine, replacement of a tool, etc. The number of activities in an AMS is large and numerous interactions are involved among these activities. The interactions exhibit concurrency, contention for resources, synchronization, randomness, and hierarchy. An AMS manager has to determine the start and finish times of various activities so that the system operation is deadlock-free and high quality output is produced at required levels (throughput) and at desired intervals (MLT).

A performance model of an AMS should be able to capture the above characteristics of an AMS. A performance model of an AMS is basically a stochastic model whose analysis would reveal the behaviour and quantitative performance of the AMS in terms of the performance measures listed in Section 2.A. There are basically two classes of such models: *simulation* and *analytical*. Simulation models can be used to construct a detailed representation of the system operation, mostly using simulation languages. However, simulation can be an expensive tool computationally, since the simulation run can be lengthy. Analytical models are more efficient than simulation if complex details of system operations are not required to be modeled. Markov chains constitute the most fundamental analytical model of AMSs. Stochastic Petri nets and queueing networks are high level Markovian models for AMSs. In [14], the authors illustrate the power of Markov models in faithfully representing a number of AMS situations such as setups, breakdowns, batching, and deadlocks, and in determining such performance measures.

Markov Models

Under suitable assumptions, an AMS can be desribed by a discrete time or continuous time Markov chain or a semi-Markov process. Such a model can naturally describe the evolution in time of an AMS. Either steady-state analysis or transient analysis could be conducted using Markov models. The presence of concurrency and synchronization in AMS interactions and the largeness of a typical AMS make the generation and solution of Markov models a formidable problem.

Stochastic Petri Net Models

Viswanadham and Narahari [16] have surveyed the use of stochastic Petri net (SPN) models in AMS performance evaluation. SPN models can capture AMS characteristics such as concurrency, randomness, and synchronization in a natural way and one can automatically generate the Markov or semi-Markov model using many available software packages. Also, using SPNs, one can investigate qualitative properties of AMSs such as those related to logical inconsistencies and instability conditions (e.g., deadlock). However the state space explosion problem has hampered the wide use of SPN models.

Queueing Network Models

Buzacott and Yao [17] have presented a survey of queueing network (QN) models of AMSs. QN models can capture resource contention, interactions, and randomness in an aggregate way, and are very attractive for steady-state analysis. A class of QNs called product form QNs have computationally efficient solution procedures and have been widely employed in modeling AMSs. QN models are useful in the preliminary design/operation stage. Also, common features such as machine blocking, assembly operations, and priority decision rules make the QN model non-product form. In such cases, only approximate analysis can be done.

C.   Reliability and Availability Measures

An AMS can be considered as an interconnected system of components such as machines, automated guided vehicles (AGVs), robots, and conveyors. These components are typically prone to *failures*. Usually, manual or automatic *repair* facilities are available, to restore failed components to the normal level of operation. Sometimes, *reconfiguration* of system resources and layout may be done, in order to cope with failures of components. Failures, together with repairs and reconfiguration, constitute three basic events that we need to model, to compute the performance of AMSs in the presence of failures.

An AMS component such as a machine or an AGV can only be in two states: up (*properly functioning*) or down (*not properly functioning*). We say a given AMS is properly functioning if the AMS is able to satisfy a given level of performance, i.e., if the throughput exceeds a given lower bound, or the MLT is below a given upper bound, the average machine utilization is above a given threshold, and so on. We will assume that such upper bounds and lower bounds of various measures with respect to which to judge the given AMS as properly functioning are given. Another way of describing an AMS that is not functioning properly, is to say that the AMS has reached a *failure state* if the performance lies out of the bounds and is *operational* if the performance is within bounds.

In this Section, we shall give several definitions to describe the behaviour of an AMS, taking into account failures, repair, and reconfiguration.

Definition 1 : *A* **fault-tolerant system** *is one that has inherent capability to automatically adapt, in a well defined manner, to failures of its components, so as to maintain continuously a specified level of performance.*

The above defines a generic fault-tolerant system and includes AMSs as a special case. When the failure of a component occurs in a fault-tolerant system, the performance level can be expected to deteriorate. However, if enough standby components are available, the same level of performance can be maintained by quickly bringing up the standby components.

Definition 2 : *A fault-tolerant system is* **degradable** *if on the occurrence of a failure, it is operable at a reduced level of performance,and* **non-degradable** *if the system continues to operate, producing the same level of performance, in the presence of component failures.*

AMSs can be classified as fault-tolerant, degradable systems.

Definition 3 : *Given an AMS, a* **structure state** *of the AMS is a vector whose components describe the condition of individual AMS subsystems as influenced by failures, repairs, and reconfigurations.*

The structure state of the system changes as time progresses. The dynamics of the state transitions is captured via the structure state process defined below.

Definition 4 : *Let $Z(u)$ be the structure state of an AMS at time $u \geq 0$. Then the family of random variables $\{Z(u) : u \geq 0\}$ is called the* **structure state process** *(SSP) of the AMS.*

The SSP of an AMS essentially characterizes the dynamics of the system structure and environmental influences, taking into account only the events concerned with failures, repairs and reconfigurations.

Example 1: Consider an AMS comprising two machines, M_1 and M_2 which are operated in parallel. When up, M_1 can produce parts at a rate of μ_1 per hour and M_2 can produce at μ_2 per hour. Assume that M_1 is faster than M_2 ($\mu_1 > \mu_2$). Let the failure times of M_i ($i = 1, 2$) be exponentially distributed with rate α_i and be independent. The structure state vector has two components here, the first indicating the status of M_1, and the second, the status of M_2. Let the state of M_i ($i = 1, 2$) be designated 1 if M_i is up, and 0 if M_i is down. Then the AMS has four structure states given by

$$S = \{(11), (10), (01), (00)\}$$

S is the state space of the structure state process. Since the up times of M_1 and M_2 are independent exponential random variables, the SSP is a continuous time Markov chain.

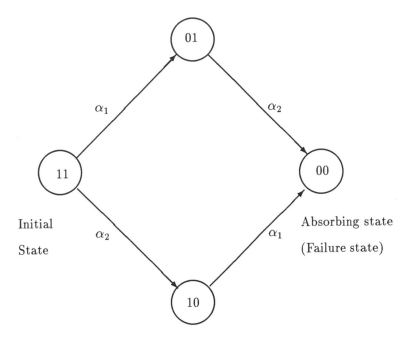

Figure 1: Structure State Process for the Non-repairable Case

Two cases would arise here, namely, the system is *non-repairable* (failed machines are not repaired) and the system is *repairable* (failed machines are repaired). In the former case, the Markov chain model for the SSP is shown in Fig. 1 (note that the state (00) is an absorbing state). Figure 2 shows the Markov SSP model for the repairable case, assuming that: (i) the repair times of M_1 and M_2 are independent exponential random variables with rates β_1 and β_2, respectively, and (ii) there is a dedicated repair facility for each machine and the repair starts immediately on failure.

Note that this AMS is fault-tolerant and degradable, with a different level of performance in each structure state.

We now define two important measures, reliability and availability.

Definition 5 : *The **reliability** of a system over an observation period* $[0, t]$ *is the probability that the system is functioning properly throughout the observation period (that is, the system does not reach a failure state during* $[0, t]$*).*

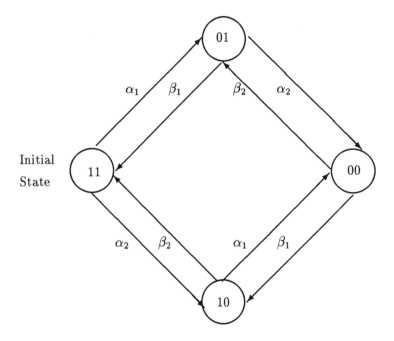

Figure 2: Structure State Process for the Repairable Case

We reiterate that the phrase "functioning properly" is with respect to a certain prespecified performance requirement. For instance, in the above example of two machines, "functioning properly" may mean "both machines working" or it may mean "at least one machine working".

Example 2: In example 1, consider the non-repairable case, where the SSP is described by the Markov chain $\{Z(u) : u \geq 0\}$ of Fig. 1. Let the state (00) be designated the system failure state. Let $Z(0) = (11)$. If the observation period is $[0, t]$, then the reliability in this case is given by

$$R(t) = P\{T > t\}$$

where T is the time to reach state (00). Note that if

$$p_{ij}(u) = P\{Z(u) = j : Z(0) = i\} \quad i, j \in S$$

give the transition probabilities of the Markov chain model, we have

$$
\begin{aligned}
R(t) &= 1 - p_{(11),(00)}(t) \\
&= p_{(11),(11)}(t) + p_{(11),(10)}(t) + p_{(11),(01)}(t)
\end{aligned}
$$

Definition 6 : *The* **instantaneous availability** $A(u)$ *of a system, at time* $u \geq 0$*, is the probability that the system is properly functioning at time* u*.*

Note that if the system is non-repairable and started in the properly functioning state, then the availability $A(u)$ and the reliability $R(u)$ are identical. However, if the system is repairable, availability and reliability will be different. For, at a given time u, the system may be functioning properly because of two mutually exclusive cases:

1. The system was properly functioning throughout the observation period $[0, u]$

2. The system was not properly functional in various subintervals of $(0, u)$, but it is properly functional at the instant u.

Availability is therefore a better measure than reliability to characterize repairable systems. Availability models usually assume that all failures are recoverable and therefore the SSP will be an irreducible Markov chain. If we assume that some failures are irrecoverable, the system will have an absorbing state (or failure state) and in such cases, we study the system reliability which will be the same as system availability. Often, we are also interested in *steady-state availability*, defined as follows:

Definition 7 : *The* **steady-state availability***, A, of a system is the limiting value of availability $A(u)$ as $u \longrightarrow \infty$. Thus,*

$$A = \lim_{u \longrightarrow \infty} A(u)$$

Note that $A = 0$ for non-repairable systems. For repairable systems, A is the fraction of time the system functions properly.

D.   Performability Measures

In this section, we present composite measures that combine both performance and reliability aspects, using the notions of structure state process and accumulated rewards. Let $\{Z(u) : u \geq 0\}$ be the SSP of an AMS. In each structure state, the system can be associated with a performance index which may be MLT, throughput, WIP, machine utilization, etc. In the most general case, the chosen performance index is a random variable. Our discussion in this paper centres around MLT and throughput because these are important and encompass other performance measures.

Definition 8 : *Given a structure state i, its associated* **reward** *f_i is a random variable that describes the performance of the system in that structure state.*

Let $[0, t]$ be an observation period. In the AMS context, this could be, for example, a shift period of 8 hours. Let $S = \{0, 1, 2, \cdots, n\}$ be the state space of the SSP $\{Z(u) : u \geq 0\}$. For $i = 0, 1, 2, \cdots, n$, define τ_i as the total time during $[0, t]$ that the SSP spends in state i. Note that

$$t = \sum_{i=0}^{n} \tau_i$$

The sum $\sum_{i=0}^{n} f_i \tau_i$ would then give the total accumulated reward during the period $[0, t]$.

Definition 9 : *Given (i) an AMS with structure state process $\{Z(u) : u \geq 0\}$ having state space $S = \{0, 1, 2, \cdots, n\}$, and (ii) rewards $f_0, f_1, \cdots, f_n$ in the individual structure states, the* **performability** $Y_t(s)$ *over an observation period $[0, t]$ and with initial structure state as $s \in S$, is a random variable given by*

$$Y_t(s) = \sum_{i=0}^{n} f_i \tau_i$$

where τ_i is the total time during $[0, t]$ the SSP stays in state i $(i = 0, 1, \cdots, n)$.

In the performability context, we will be interested in three measures: performability distribution, steady-state performability, and interval performability. These are defined below.

Definition 10 : *The* **performability distribution** *is the cumulative distribution function of performability $Y_t(s)$, i.e., $P\{Y_t(s) \leq x\}$ for $x \in \Re$; the limit*

$$\lim_{t \longrightarrow \infty} Y_t(s)$$

if it exists, is called the **steady state performability;** *and the expected value $E[Y_t(s)]$ is called the* **interval performability.**

Example 3: Consider the two machine system of example 1 and assume that repairs are not possible. The SSP $\{Z(u) : u \geq 0\}$ of the system is given in Fig. 1. Let us relabel the state (00) as state 0, (01) as 1, (10) as 2, and (11) as 3. The state space S of the SSP is then given by $S = \{0, 1, 2, 3\}$. Assume that 3 is the initial state.

First we look at throughput-related performability. For this, we choose reward f_i $(i = 0, 1, 2, 3)$ as the average throughput rate in state i. Assuming that raw parts are always available, we then have

$$f_0 = 0; \quad f_1 = \mu_2; \quad f_2 = \mu_1; \quad f_3 = \mu_1 + \mu_2$$

In an observation period $[0, t]$, let τ_i as usual denote the time spent in state i. We then see that the sum $\tau_1 \mu_2 + \tau_2 \mu_1 + \tau_3(\mu_1 + \mu_2)$ is the total accumulated average throughput during $[0, t]$. This will be denoted by $Y_t(3)$. By computing the distribution functions of τ_1, τ_2, and τ_3, one can evaluate, for any throughput lower bound x, the probability $P\{Y_t(3) > x\}$, of exceeding a throughput of x during $[0, t]$. Now, $E[Y_t(3)]$ is the interval performability or average throughput.

Next, we look at MLT-related performability. Note that average MLT in state 1 is $W_1 = 1/\mu_2$, average MLT in state 2 is $W_2 = 1/\mu_1$, and in state 3 the average MLT is $W_3 = 1/(\mu_1 + \mu_2)$. Since no parts are produced in state 0, it does not affect the analysis. Note that τ_1, τ_2, and τ_3 are random variables and so will be $Y_t(3)$.

Suppose D is a desired upper bound on MLT and we choose the rewards as

$$f_0 = 0\,; \quad f_i = (W_i - D)Q_i \quad (i = 1, 2, 3)$$

where W_i is the average MLT and Q_i is the average throughput in state i. The accumulated reward is

$$Y_t(3) = \sum_{i=0}^{3} \tau_i (W_i - D)Q_i$$

Now consider

$$
\begin{aligned}
P\{Y_t(3) < 0\} &= P\{\sum_{i=0}^{3} \tau_i (W_i - D)Q_i < 0\} \\
&= P\{\sum_{i=0}^{3} \tau_i W_i Q_i < \sum_{i=0}^{3} \tau_i D Q_i\} \\
&= P\{\frac{\sum_{i=0}^{3} \tau_i W_i Q_i}{\sum_{i=0}^{3} \tau_i Q_i} < D\}
\end{aligned}
$$

The above probability can be interpreted as the probability of the overall average MLT during $[0, t]$ being less than a desired upper bound D.

Thus, by appropriately defining the rewards, one can compute performability measures of various kinds. The above discussion can be extended to the case when the SSP is as described in Fig. 2.

3   Performability of a Versatile Machine Center

In this section, we consider an AMS comprising a single NC machine center consisting of one or more identical, failure-prone NC machines. The architecture of such an NC machine center is shown in Fig. 3. A typical system incorporates the following subsystems:

1. A *pallet pool* of fixtured raw parts awaiting processing.

2. One or more *NC machines* for processing semi-finished or raw parts.

3. A *tool magazine* and an *automatic tool changer* that will load and unload required tools as and when necessary.

4. A *rotary indexing buffer turn table* for simultaneous loading and unloading of a finished part.

5. A *transportation mechanism*, in the form of an automated guided vehicle (AGV) to transport raw and finished parts between the pallet pool and unload area

6. A *programmable logic controller* to supervise all operations in the system.

Typically, the machine centers are versatile and can produce parts of several different part types by executing appropriate part programs and employing proper tools.

We shall illustrate the performability concept for the single part type case. Generalization to the multiple part type case can be easily carried out. Also, we will assume that the center has only one machine. This is done for simplicity in presentation.

Two natural and possibly the simplest queueing models that are applicable to the present problem are:

1. M/M/1/∞ queue with breakdowns

2. M/M/1/N queue with breakdowns

For both the above models, the SSP has only two states: machine in working condition ("up") and machine in failed condition ("down"). Assuming that the time to next failure and the time to repair are independent exponential random variables, this SSP will be a Markov chain. Figure 4(a) shows the state diagram of the SSP assuming the machine is repairable, and

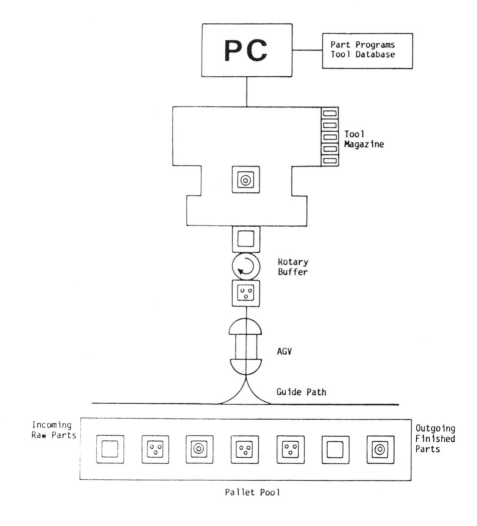

Figure 3: Machine Center with Pallet Pool

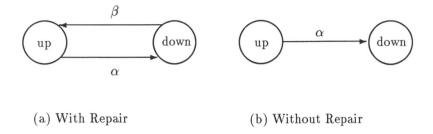

(a) With Repair   (b) Without Repair

Figure 4: State Diagram of the Structure State Process of a Single Machine System

Fig. 4(b) shows the SSP for the unrepairable case. In the latter case, the down state will be an absorbing state.

The steady-state analysis of a single server queueing system with breakdowns has been an active topic of research interest. Doshi [18] has surveyed recent results on this topic. Our discussion in this paper will be based on the papers by Gaver [19], Mitrani and Avi-Itzhak [20], and Bobbio and Trivedi [21].

A.   Machine Center with Infinite Waiting Space

The model here is an $M/M/1/\infty$ queue with infinite queue and an unreliable server. Assume that raw parts arrive into the system according to a Poisson process with rate λ and the service times of parts are i.i.d. exponential random variables with mean $1/\mu$. The machine is assumed to break down such that the time between successive failures is an exponential random variable with mean $1/\alpha$. The service is interrupted immediately at a failure and simultaneously the repair commences. Repair times are assumed to be i.i.d. exponential random variables with mean $1/\beta$. As soon as repair finishes, the machine resumes processing on the interrupted workpiece. To complete the model description, we additionally assume the following:

1. The machine can fail when it is idle or when it is processing a part.

2. When the machine is under repair, no further failures can occur.

For this system, we now compute two steady-state performability measures, namely average MLT and average WIP. To this end, we first write down the rate balance equations. Figure 5 shows the state diagram of this queueing system. Let $X(u)$ be the state of the machine at time u and $Y(u)$, the number of parts in the system at time u. We see that the stochastic process $\{(X(u), Y(u)) : u \geq 0\}$ constitutes a continuous time Markov chain. Let '0' indicate the machine down condition, and a '1' indicate the machine is up. The state space of the process is

$$S = \{(i, j) : i \in \{0, 1\}, j \in \{0, 1, 2, \cdots\}\}$$

Let $\pi(i, j)$ be the steady-state probability of state (i, j). The rate balance equations are given by

$$
\begin{aligned}
(\lambda + i\mu + i\alpha + (1 - i)\beta)\pi(i, k) = \ & \lambda\pi(i, k - 1) + \\
& i\mu\pi(i, k + 1) + \\
& i\beta\pi(1 - i, k) + (1 - i)\alpha\pi(1 - i, k) \\
& \text{for } k \geq 1 \tag{1}
\end{aligned}
$$

$$
\begin{aligned}
(\lambda + i\alpha + (1 - i)\beta)\pi(i, 0) = \ & i\mu\pi(i, 1) + i\beta\pi(1 - i, 0) \\
& + (1 - i)\alpha\pi(1 - i, 0) \tag{2}
\end{aligned}
$$

where $i = 0, 1$ and $k = 0, 1, 2, \cdots$. The solution of the above system of equations may be carried out by employing the method of generating functions, as given by Mitrani and Avi-Itzhak [20]. It can be shown that the mean WIP (denoted L) in the steady-state is given by

$$L = \frac{\lambda[(\alpha + \beta)^2 + \mu\alpha]}{(\alpha + \beta)[\beta(\mu - \lambda) - \lambda\alpha]} \tag{3}$$

Little's law can be invoked to compute W, the mean MLT in the steady-state:

$$W = \frac{L}{\lambda} = \frac{(\alpha + \beta)^2 + \mu\alpha}{(\alpha + \beta)[\beta(\mu - \lambda) - \lambda\alpha]} \tag{4}$$

The condition for the stability of this queueing system is given by

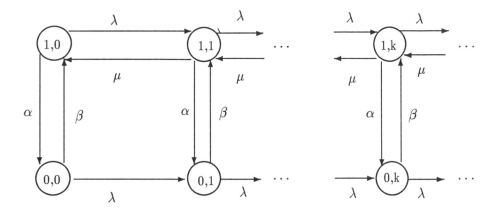

Figure 5: State Diagram for $M/M/1/\infty$ Queue with Breakdowns

$$\frac{\lambda}{\mu} < \frac{1}{1 + \frac{\alpha}{\beta}} \qquad (5)$$

or

$$\frac{1}{\lambda} > \frac{1}{\mu}(1 + \frac{\alpha}{\beta}) \qquad (6)$$

The term $1/(1+\alpha/\beta)$ is the steady-state probability of the machine being "up" and is therefore the *steady-state availability* of the system. Further, the quantity $(1/\mu)(1 + \alpha/\beta)$ will give the *mean completion time* of a part.

For the above system, performance measures of the associated $M/M/1/\infty$ queue without breakdowns would give the "performance" related measures whereas steady-state availability would give a "reliability" measure. The mean completion time, mean steady-state WIP as given by Eq. (3), and mean steady-state MLT as given by Eq. (4) would give performability measures.

Example 4: Consider a machine center with one machine that can produce 20 parts per hour (mean processing time = 3 minutes). Fixtured raw parts are found to arrive according to a Poisson process with rate 10 per hour (average of 6 minutes for inter-arrival time). The machine is assumed to fail randomly and the average repair time is assumed to be 1 hour. We can model the above system as a $M/M/1/\infty$ queue with breakdowns. Figure 6 shows the variation of mean steady-state MLT as a function of the failure rate α (the average WIP is linearly

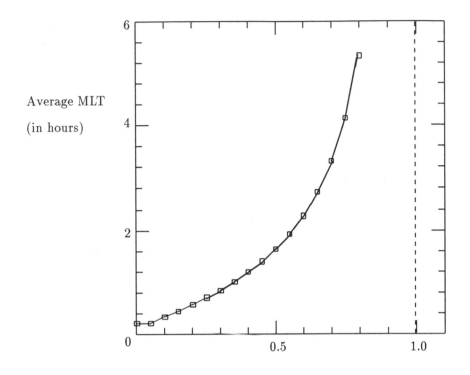

Figure 6: Variation of Average MLT with Failure Rate

related to mean MLT, and hence behaves similarly). For a $M/M/1/\infty$ *without* failures, the mean MLT would be 6 minutes and the mean WIP would be 1. It is observed from the graph in Fig. 6 that the MLT (as well as WIP) tends to as ∞ as α approaches the value of 1 per hour. The graph shows that inventory levels and lead times can rise rapidly if the system is prone to increasing rate of failures.

B.   Machine Center with Finite Waiting Space

In the AMS context, finite waiting space is more realisitic than the previously considered case of infinite waiting space, as exemplified by the pallet pool in Fig. 3. An appropriate model here is an M/M/1/N queue with

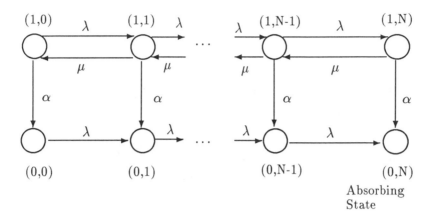

Figure 7: State Diagram for the Markov chain model of $M/M/1/N$ Queue with Breakdowns and No Repair

breakdowns, where N is the total number of parts that can be accommodated in the system. Following the same notation as in the previous section, the state space of the Markov chain is given by

$$\{(i,j) : i = 0, 1; j = 0, 1, 2, \cdots N\}$$

Here, we assume that once the machine fails, it is not repairable. The model here is a finite Markov chain, whose state transition graph is shown in Fig. 7.

Bobbio and Trivedi [21] have shown for the above queueing system that the average number of jobs $Q(t)$ completed in a finite observation period $[0, t]$ is given by

$$Q(t) = \frac{\mu}{\alpha} \frac{(1 - e^{-\alpha t})\rho(1 - \rho^N)}{(1 - \rho^{N+1}) + \frac{\alpha}{\lambda}e^{-\alpha t}[N - \frac{\rho}{1-\rho}(1 - \rho^N)]} \tag{7}$$

where $\rho = \lambda/\mu$.

The above is the *total average throughput* of the system during the interval $[0, t]$ and hence is the interval performability measure for throughput. In the limit, as t approaches ∞, we obtain the average number of jobs completed before system failure, $\overline{Q}$, as

$$\overline{Q} = \frac{\mu\rho(1 - \rho^N)}{\alpha(1 - \rho^{N+1})}$$

Example 5: It would be interesting to study the total average throughput accumulated over different time intervals for different failure rates of the machine. Let $\lambda = 10$ parts per hour, and $\mu = 20$ per hour. Assume that at most 5 parts can be inside the system at any time, i.e., $N = 5$. Figure 8 shows a graph of accumulated average throughput values over different time interals. The failure rates considered are $\alpha = 0.01, 0.1, 0.2, 0.5$ per hour. In all the cases, it is seen that the throughput varies almost linearly in the initial stages and then saturates. The saturation occurs at smaller t for larger values of α. This is reasonable to expect since as soon as a failure occurs, the system becomes unproductive as there is no repair. Thus given the failure characteristics of a machine center, it would be possible to say whether or not a given average throughput can be delivered in a given observation period. The effect of α on the cumulative throughput is quite significant and therefore the predicted values are very different from those of the failure-free case.

4   AMS with Multiple Machines and Centralized Material Handling

In the previous section, we have illustrated the concepts of steady-state performability and interval performability in the AMS context. In this section, we look at the computation of the distributions of performability measures using the notions of structure state process and accumulated rewards.

The example that we consider is that of an AMS with K machine centers $M_1, M_2, \cdots, M_K$, and a pool of automated guided vehicles (AGVs) for material handling. A closed queueing network model of this system under failure-free conditions is shown in Fig. 9, assuming a single machine in each machine center and a single AGV in the AGV pool. This model is called the *closed central server model* and we shall refer to the AMS as the central-server AMS. This model is a well-known and oft-employed model for flexible manufacturing systems (FMSs), first formulated by Solberg [22]. The parameters in this model are: K, the number of machine centers; m_0, the number of AGVs; the number of machines m_i in machine

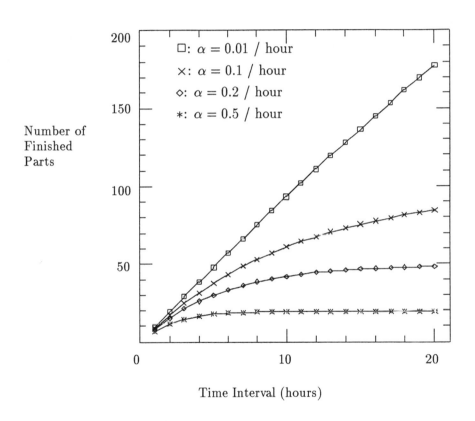

Figure 8: Interval Performability for a Machine Center with a Non-repairable Machine and a Finite Pallet Pool

center i ($1 \leq i \leq K$); $q_0, q_1, \cdots, q_K$, the routing probabilities, which satisfy $\sum_{i=0}^{K} q_i = 1$; $\mu_0, \mu_1, \cdots, \mu_K$ (service rates of the nodes); and N (population of jobs inside the network).

The closed nature of the model means that the population of jobs remains the same inside the FMS. This captures in a natural way the fact that the number of pallets or fixtures in an FMS is constant. A fixtured raw part is loaded into the system by the AGV and is routed to one of the machine centers with probabilities $q_1, q_2, \cdots, q_K$. Each machine center will contain a certain number of individual machines, which are fed from the common pre-process buffer queue of waiting parts in front of this machine center. After processing finishes in a machine center, the AGV unloads the part from the system if all operations are over, or carries it to another machine center for the next operation. When a finished part is unloaded from the system, the fixture or pallet held by the part is released. The released fixture or pallet can be utilized for the next raw part. It is assumed that raw parts are always available. The arc labeled by q_0 indicates the exit of a finished part and also the entry of a new raw part. Simple queueing analysis of this model shows that each part undergoes on an average $1/q_0$ operations and visits machine center M_i, q_i/q_0 times on the average. Using standard computational algorithms for closed queueing networks, one can compute efficiently the different performance measures for this model, such as MLT and throughput.

A.   Central Server FMS without Repair

We shall consider a special case of a central server FMS as shown in Fig. 10. Here, we have one AGV for material handling ($m_0 = 1$), and there is a single machine center (i.e., $K = 1$) with m_1 identical machines in it. Since we have only one machine center, we drop the suffix from m_1 in the following discussion. If the AGV and the machines are prone to failure, then the structure state process $\{Z(u) : u \geq 0\}$ for this system will have the state space $\{0, 1, \cdots, m\}$ where the interpretation of the states is as follows:

    0 : AGV failed or all machines failed

    i : AGV operating and exactly i machines working ($i = 1, 2, \cdots, m$)

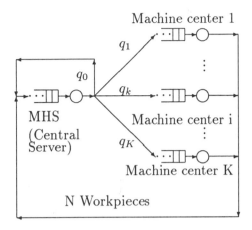

Figure 9: The Central Server Model of an FMS

Let the time to failure of the AGV and the time to failure of each machine be exponentially distributed with rates α_A and α respectively, and let the failures be independent of one another. Let us assume that a device (AGV or machine), once failed, is not repaired in the observation period. In such a case, the state transition graph of the SSP is as shown in Fig. 11.

The SSP depicted in Fig. 11 is the same as the one considered by Donatiello and Iyer [6]. We will follow the same approach as presented by Donatiello and Iyer to compute the performability distribution of this FMS. The scenario we consider is the following: the FMS in question has an AGV and m machines which are failure prone. The system has N fixtures. At the end of an observation period, say a shift period of 8 hours, the FMS is to meet a minimum production target and also produce the parts in as short a time as possible. Let the observation period be $[0, t]$, the total production during this period $Q(t)$, and the MLT of a typical part, $W(t)$. Also, assume that x is a given production target, w a desired MLT, and p a desired probability. We would like to answer questions such as:

1. For a given m, x, w, and p, how many fixtures should we employ, i.e, what should be the value of N, in order that

$$P\{Q(t) > x\} > p \tag{8}$$

$$P\{W(t) < w\} > p \tag{9}$$

2. For a given N, x, w and p, many machines (m) should we employ so as to satisfy Eq. (8) or Eq. (9) ?

3. What are the values of $E[Q(t)]$ and $E[W(t)]$?

4. For a given m, x, w, p, and N, what should be t (how long should the shift be operated) so that Eq. (8) is satisfied ?

All of the above questions can be answered by computing over $[0, t]$ the throughput-related and MLT-related performability distributions. To achieve this, we proceed as follows.

Consider the SSP $\{Z(u) : u \geq 0\}$ of Fig. 11. Let f_i be the reward associated with state i ($i = 0, 1, \cdots, m$). Assume that the initial state is m (i.e., all the machines and the AGV are functioning). In the observation period $[0, t]$, the accumulated reward or the performability is given by

$$Y_t(m) = \sum_{i=0}^{m} f_i \tau_i$$

where τ_i is the total time during $[0, t]$ the SSP stays in state i. Starting from state m, the evolution of SSP during $[0, t]$ can fall into three different categories.

1. The SSP stays in state m throughout the interval, without making a transition to any other state.

2. The SSP transits to state 0 directly from state m, some time during the interval and will therefore stay in state 0 for the rest of the interval.

3. The SSP transits to state $m-1$ at some instant during the observation period. Its evolution during the rest of the interval will follow the same pattern as the original process, except that the initial state will be $m - 1$.

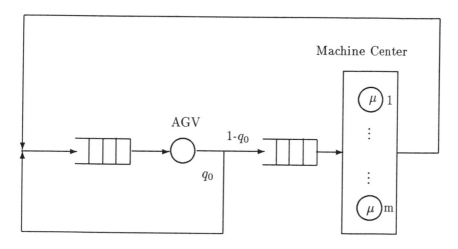

Figure 10: Central Server Model of an FMS with an AGV and a Machine Center with m Identical Machines

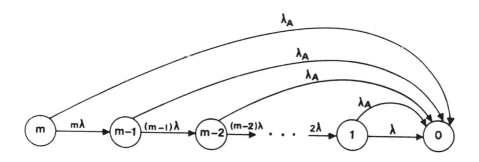

Figure 11: Structure State Process for the Central Server FMS Without Repair

The above observations form the basis of a recursive formulation for the evolution of the SSP over $[0, t]$. To write down the recursive formulation, we first introduce some notation. Let C be a general condition. Define the indicator function

$$I(C) = \begin{cases} 1 & \text{if } C \text{ is true} \\ 0 & \text{if } C \text{ is false} \end{cases}$$

Let p_{ij} $(i = 0, 1, \cdots, m; \ j \leq i)$ denote the probability of a single-step transition from state i to state j. Note that, for $k = 1, 2, \cdots, m$,

$$p_{k,0} = \frac{\alpha_A}{\alpha_A + k\alpha}$$

$$p_{k,k-1} = \frac{k\alpha}{\alpha_A + k\alpha}$$

For $k = 0, 1, \cdots, m$, let $c_k = \alpha_A + k\alpha$. It may be verified that the sojourn time in state $k \geq 1$ is exponentially distributed with rate c_k. Therefore, $e^{-c_m t}$ gives the probability that the SSP stays in the initial state m throughout the interval $[0, t]$.

Consider the probability $P\{Y_t(m) < y\}$ where $y > 0$. This can be split into three terms, based on the three observations above. Donatiello and Iyer [6] have shown, for $m > 1$, that

$$\begin{aligned} P\{Y_t(m) < y\} \ = \ & I(f_m t < y) e^{-c_m t} \\ & + \int_0^t c_m e^{-c_m x} p_{m0} I([f_m x + f_0(t - x)] < y) dx \\ & + \int_0^t c_m e^{-c_m x} p_{m,m-1} P\{Y_{t-x}(m - 1) < y - f_m x\} dx \end{aligned}$$

$$(10)$$

Equation (10) gives a recursive formulation for computing the distribution of performability. For $m = 1$, this equation applies without the first term in the two integrals and after setting $p_{m,m-1} = 1$. Also, $Y_t(0) = f_0 t$, and $c_0 = 0$ since state 0 is an absorbing state of the SSP. The solution of Eq. (10), proposed by Donatiello and Iyer [6], uses the double Laplace-Stieltjes transform of the distribution $P\{Y_t(m) < y\}$ to unfold the recursion into an efficient computational procedure. We do not present the details of this procedure, but only give some interesting numerical results obtained by us using the software implementation of the above procedure.

Example 6: *Throughput-related performability*

First, we look at a central server FMS with number of fixtures, $N = 12$. Let the observation period be a shift of 8 hours. The other parameters are:

Number of machines $= m$ (variable).

The exit probability of finished part leaving the system (with the simultaneous entry of a new part) is $q_0 = 0.2$, implying that a workpiece undergoes five material handling operations and four machining operations on the average.

The rate of material movement by the AGV (μ_0) is assumed to be 100 parts per hour, and the mean processing rate of a machine (μ_1) is 20 parts per hour.

Mean time to failure (MTTF) of AGV $= 8$ hours.

MTTF of each machine $= 16$ hours.

Table I shows the probability that the number of finished parts produced in the shift period exceeds x, for $x = 20, 30, 40, 50, 60,$ and, 70, for $m = 1, 2, \cdots, 8$. Using this table, we can answer questions such as, 'what is the initial number of "up" machines so that the probability of the total production in the shift exceeding a given number is greater than a given probability ?'. For example, if it is required that this probability should be greater than 0.75, then we need 3 machines to surpass 20 parts (and also 30 parts); 5 machines to surpass a target of 40 parts, and so on.

Figure 12 shows, for three different shift durations, 4 hours, 6 hours, and 8 hours, the *dual* of the distribution function of the cumulative production during the shift. It is assumed that the number of machines is 4, and that the number of fixtures is 12. The other parameters remain as above.

Example 7: *MLT-related performability*

We now consider MLT-related performability for the central server FMS. Let the shift duration be 8 hours, $m = 4$, $1/\alpha_A = 8$ hours, $1/\alpha = 16$ hours. (Other parameters are the same as in Example 6). For different lower bounds on MLT, and for different numbers of fixtures, Table II shows the probability

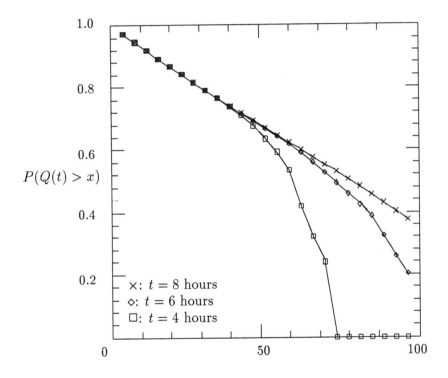

Target Cumulative production, x

Figure 12: Dual of the Distribution Function of the Cumulative Production in Example 6

Table I: The impact of number of machines on meeting target production levels

No. of. machines	Probability of manufacturing a specified no. of parts Prob$\{Q(t) > x\}$					
	Target production level, x					
	20.0	30.0	40.0	50.0	60.0	70.0
1	0.4724	0.3247	0.0000	0.0000	0.0000	0.0000
2	0.7407	0.6200	0.5126	0.3760	0.2724	0.1941
3	0.8338	0.7525	0.6738	0.5942	0.5137	0.4364
4	0.8675	0.8043	0.7429	0.6831	0.6240	0.5655
5	0.8781	0.8214	0.7671	0.7151	0.6650	0.6163
6	0.8811	0.8265	0.7748	0.7258	0.6792	0.6347
7	0.8819	0.8281	0.7773	0.7294	0.6841	0.6413
8	0.8822	0.8286	0.7781	0.7306	0.6859	0.6438

values of attaining the lower bounds on MLT. This table can be used to answer the question "what is the maximum number of fixtures that may be employed so that the probability of the MLT being less than a specified value exceeds a desired probability ?". For example, if the desired probability is 0.95, then we can employ at most 4 fixtures for a lower bound of 60 minutes, at most 13 fixtures for a lower bound of 90 minutes and so on.

Figure 13 shows a graph of the cumulative distribution function for MLT-related performability, assuming $m = 4$ and $N = 12$, other parameters remaining the same as above. Three possible shift durations are considered, namely 4 hours, 6 hours and 8 hours.

B. Central Server FMS with Repairs

The system considered is the same as in Section 4.A, except that the AGV and the machines are now repairable. We assume that we have a single repair facility before which failed devices queue up for repair. There are two types of customers for the repair facility: AGV and machines. Because of the importance of the AGV for sustaining production in the system, we assume that the repair facility will attend to the AGV as soon as the AGV fails, even if a machine is under repair at that time. The AGV repair is

Table II: The effect of the number of fixtures in attaining a desired level of MLT

Probability that MLT is less than a specified value $\text{Prob}\{W(t) < w\}$							
No. of. fixtures	Desired MLT, w						
	25.00	35.00	45.00	55.00	65.00	75.00	85.00
4	0.8112	0.9820	0.9978	1.0000	1.0000	1.0000	1.0000
5	0.4377	0.9425	0.9887	0.9979	0.9998	1.0000	1.0000
6	0.0000	0.8428	0.9713	0.9915	0.9979	0.9997	1.0000
7	0.0000	0.6432	0.9214	0.9796	0.9929	0.9979	0.9995
8	0.0000	0.4022	0.8438	0.9549	0.9841	0.9938	0.9979
9	0.0000	0.0000	0.7240	0.9046	0.9699	0.9869	0.9944
10	0.0000	0.0000	0.5113	0.8409	0.9378	0.9768	0.9887
11	0.0000	0.0000	0.3681	0.7611	0.8916	0.9567	0.9807
12	0.0000	0.0000	0.0000	0.5788	0.8375	0.9233	0.9678
13	0.0000	0.0000	0.0000	0.4457	0.7794	0.8813	0.9437
14	0.0000	0.0000	0.0000	0.0000	0.6228	0.8344	0.9112
15	0.0000	0.0000	0.0000	0.0000	0.5019	0.7848	0.8731

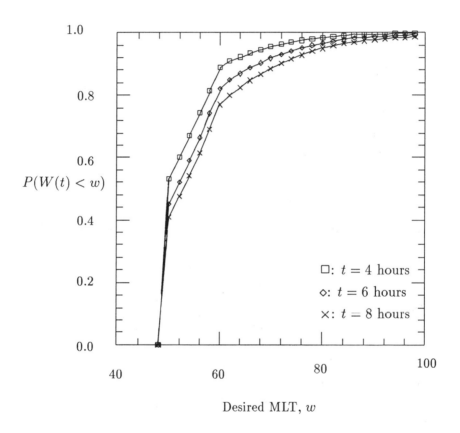

Figure 13: Cumulative Distribution Function of the Manufacturing Lead Time in Example 7

not preemptable and this repair time is exponential with rate β_A. The repair of a machine can be preempted in the event of a AGV failure, and if preempted, it is assumed that the repair of the interrupted machine will resume immediately when the repair of the AGV is completed. Machine repair time is exponential with rate β. The machines that have failed are assumed to wait in FCFS fashion for the repair operation. The AGV and the machines fail independently of one another with failure rates α_A and α respectively, and they can fail even when idle. The repair facility does not fail, and the devices being repaired do not fail during the repair.

Figure 14 shows a closed queueing network model with two customer classes for the failure-repair behaviour described above. It is interesting to note that this is just the *machine repairman model* with two job classes. The SSP is the Markov chain underlying this model and Fig. 15 shows the state transition graph of this SSP. Note that the SSP is an irreducible and positive recurrent Markov chain, in contrast with the acyclic chain of Fig. 11.

Computation of distributions of performability when the SSP is a non-acyclic Markov chain has received attention only recently. Silva and Gail [10] have proposed a computational approach, based on coloring subintervals of observation period, to compute the performability distributions. Their method uses the randomization technique. Kulkarni *et al* [23] describe an algorithm for numerical evaluation of the performability distributions in repairable systems. Their solution is in terms of the Laplace-Stieltjes transform of the distributions. Iyer *et al* [8] have presented a computational method for computing moments of performability for repairable systems. Smith *et al* [9] have come up with an efficient polynomial-time algorithm for numerical evaluation of performability distributions in repairable systems.

Example 8: Table III shows a comparison of the throughput levels that can be obtained in the repairable and the non-repairable cases for different time durations. The parameters assumed are the following: $m = 4$; $N = 12$; $1/\alpha_A =$ 8 hours; $1/\alpha = 16$ hours; $1/\beta_A = 2$ hours; $1/\beta = 4$ hours. (Other parameters are identical to those in Example 6). For each case, the table shows the expected throughput at the end of the observation period (interval performability). Recall that the rewards in this case are obtained using the central server model. The relevant quantities of interest for the structure state model have been obtained

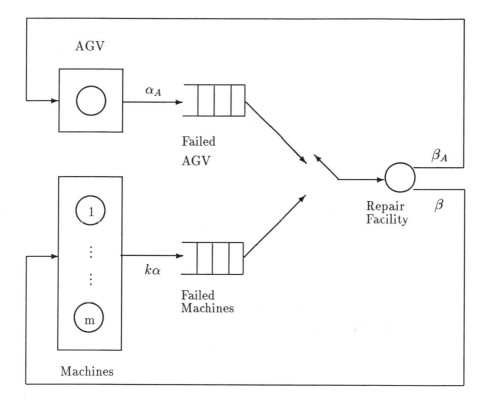

Figure 14: Structure State Model for the Central Server FMS with Repair

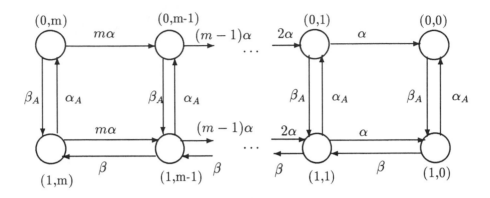

Figure 15: State Transition Diagram of the Structure State Process of the Machine Repairman Model of Fig. 14

using the technique of Silva and Gail [10].

5   Conclusions

Automated manuafacturing systems of today are endowed with character-istics such as flexibility, fault-tolerance, and degradability in performance. Neither pure reliability analysis nor pure performance analysis can accu-rately capture the behaviour of such systems.   Integrated performance-reliability or *performability* studies therefore assume great importance.

In this contribution, we have shown, through illustrative examples of manufacturing systems, the utmost relevance of performability modeling in AMS design. The performability measures we have considered deal with throughput and manufacturing lead time, which essentially decide the com-petitiveness of a plant.

There is a rich body of literature on the topic of performability modeling in fault-tolerant computing and this topic remains an active area of research today. In our view, this paper has opened up a new area and it would be interesting to see how the available literature on performability can benefit studies related to AMSs.

Table III: Cumulative production over various shift periods: Repairable versus non-repairable case

Mean number of parts produced $E[Q(t)]$		
Shift duration, $[0, t]$ (hours)	Non-repairable case	Repairable case
4.0	52.4410	59.1739
5.0	60.6821	71.8771
6.0	67.5594	84.1832
7.0	73.2929	96.1694
8.0	78.0685	107.8877
9.0	82.0434	119.3759
10.0	85.3498	130.6631
11.0	88.0989	141.7730
12.0	90.3835	152.7253
13.0	92.2814	163.5373
14.0	93.8576	174.2237
15.0	95.1663	184.7977
16.0	96.2527	195.2709

References

[1] M. Beaudry, "Performance related reliability measures for computer systems," *IEEE Transactions on Computers*, vol. 27, pp. 248–255, June 1978.

[2] J. F. Meyer, "On evaluating the performability of degradable computing systems," *IEEE Transactions on Computers*, vol. 29, pp. 720–731, August 1980.

[3] J. F. Meyer, "Closed-form solutions of perfomability," *IEEE Transactions on Computers*, vol. 31, pp. 648–657, July 1982.

[4] A. L. Reibman, "Modeling the effect of reliability on performance," *IEEE Transactions on Reliability*, vol. 39, pp. 314–320, August 1990.

[5] A. Goyal, S. S. Lavenberg, and K. S. Trivedi, "Probabilistic modeling of computer system availability," *Annals of Operations Research*, vol. 8, pp. 285–306, 1987.

[6] L. Donatiello and B. R. Iyer, "Analysis of a composite performance reliability measure for fault-tolerant systems," *Journal of the Association for Computing Machinery*, vol. 34, pp. 179–199, January 1987.

[7] V. F. Nicola, V. G. Kulkarni, and K. S. Trivedi, "Queueing analysis of fault-tolerant computer systems," *IEEE Transactions on Software Engineering*, vol. 13, pp. 363–375, March 1987.

[8] B. R. Iyer, L. Donatiello, and P. Heidelberger, "Analysis of performability for stochastic models of fault-tolerant systems," *IEEE Transactions on Computers*, vol. 35, pp. 902–907, October 1986.

[9] R. M. Smith, K. S. Trivedi, and A. V. Ramesh, "Performability analysis: Measures, an algorithm and a case study," *IEEE Transactions on Computers*, vol. 37, pp. 406–417, April 1988.

[10] E. de Souza e Silva and H. R. Gail, "Calculating availability and performability measures of repairable computer systems using randomisation," *Journal of the Association for Computing Machinery*, vol. 36, pp. 171–193, January 1989.

[11] G. Ciardo, R. A. Marie, B. Sericola, and K. S. Trivedi, "Performability analysis using semi-Markov reward processes," *IEEE Transactions on Computers*, vol. 39, pp. 1251–1264, October 1990.

[12] M. L. Kanth and N. Viswanadham, "Reliability analysis of flexible manufacturing systems," *International Journal of Flexible Manufacturing Systems*, vol. 2, pp. 145–162, 1989.

[13] V. Albino, G. O. Okogbaa, and R. L. Shell, "Computerized integrated performance–reliability measure of a flexible automated production system," *Computers and Industrial Engineering*, vol. 18, no. 4, pp. 547–558, 1990.

[14] N. Viswanadham and Y. Narahari, "*Performance Modeling of Automated Manufacturing Systems.*" To be published by Prentice-Hall, Inc., Englewood Cliffs, New Jersey, USA, 1992.

[15] Y. C. Ho (Ed.), "Dynamics of discrete event systems." Special Issue of the *Proceedings of the IEEE*, vol. 77, no. 1, January 1989.

[16] N. Viswanadham and Y. Narahari, "Stochastic Petri net models for performance evaluation of automated manufacturing systems," *Information and Decision Technologies*, vol. 14, pp. 124–142, 1988.

[17] J. A. Buzacott and D. D. Yao, "On queueing network models of flexible manufacturing systems," *Queueing Systems*, vol. 1, pp. 5–27, 1986.

[18] B. Doshi, "Single server queues with vacations," in *Stochastic Analysis of Computer and Communication Systems* (H. Takagi, ed.), pp. 217–265, Amsterdam: North Holland, 1990.

[19] D. P. Gaver, "A waiting line with interrupted service, including priorities," *Journal of the Royal Statistical Society, Series B*, vol. 24, pp. 73–90, 1962.

[20] I. L. Mitrani and B. Avi-Itzhak, "A many-server queue with service interruptions," *Operations Research*, vol. 16, pp. 628–638, 1968.

[21] A. Bobbio and K. S. Trivedi, "An aggregation technique for the transient analysis of stiff Markov chains," *IEEE Transactions on Computers*, vol. 35, pp. 803–814, September 1986.

[22] J. J. Solberg, "A mathematical model of computerized manufacturing systems," in *Proceedings of the Fourth International Conference on Production Research,* Tokyo, pp. 1265–1275, 1977.

[23] V. G. Kulkarni, V. F. Nicola, R. M. Smith, and K. S. Trivedi, "Numerical evaluation of performability and job completion time in repairable fault-tolerant systems," in *Proc of the Symposium on Fault-Tolerant Computing, Vienna, Austria,* pp. 252–257, IEEE, 1986.

MODELING, CONTROL, AND PERFORMANCE ANALYSIS OF AUTOMATED MANUFACTURING SYSTEMS USING PETRI NETS

FRANK DICESARE
Electrical, Computer, and Systems Engineering Department
Center for Manufacturing Productivity and Technology Transfer
Rensselaer Polytechnic Institute
Troy, New York 12180-3590

ALAN A. DESROCHERS
Electrical, Computer, and Systems Engineering Department
Center for Intelligent Robotic Systems for Space Exploration
Rensselaer Polytechnic Institute
Troy, New York 12180-3590

I. INTRODUCTION

Consider an automated manufacturing system. The goal is to make a set of products from raw materials and purchased parts using resources such as machines, robots, materials handling and storage devices. A complex set of activities must occur in order to meet this goal. The need exists in manufacturing to properly coordinate and synchronize these activities and resources which work concurrently to produce a set of products. This is the manufacturing control design problem.

This chapter emphasizes the Petri net approach for modeling, control, and performance analysis of automated manufacturing systems. This approach has become more important in recent years because it can solve problems that cannot be modeled using queueing theory, while avoiding the time consuming, trial and error approach of simulation.

The modeling problem is characterized by concurrent and asynchronous events that are typical for such discrete event dynamic systems. Petri nets are well suited for modeling manufacturing systems because they capture the precedence relations and interactions among these events. In addition, a strong mathematical foundation exists for describing these nets. This allows a qualitative analysis of such system properties as deadlock, conflict, and boundedness.

The Petri net model can also be used as the basis of a real-time controller for a manufacturing system. The flow of tokens through the net establishes the sequence of events to carry out a specific task, such as the manufacturing of a particular part type. Petri net controllers have been used in factories in Japan and Europe.

Petri nets are also a very valuable performance analysis tool. When time is added to the firing of the transitions, it becomes possible to calculate temporal measures that analyze the merit of a particular production system. If time is allowed to be a random variable described by an exponential distribution, then these nets are referred to as stochastic Petri nets. It can be shown that these nets are equivalent to a Markov chain, but much more compact in the representation of the system. The compactness allows for the handling of systems with a large number of states (on the order of thousands) and the equivalence with Markov chains permits the calculation of performance measures like throughput, average machine utilization, probabilitiy that a machine is blocked or starved, etc. Examples from transfer lines and production networks are included.

What are Petri nets?

Petri net theory was originally developed by Carl Adam Petri and presented in his doctoral dissertation in 1962. Petri nets are a *graph theoretic* as well as a *visually graphical* tool specifically designed for modeling, analysis, performance evaluation and control of interacting concurrent discrete event systems. These types of systems are characterized by a discrete state space and are event driven, as opposed to time dependent. They are sometimes called discrete event systems (DES) or discrete event dynamic systems (DEDS). Examples of DES are manufacturing, communication networks, and computers.

The following will introduce Petri nets informally using the simple robotic example in Figure 1. In this example, we want to show the sequence of conditions and events that must occur for the robot to move two parts. We can describe a Petri net (PN) graphically as having several elements:

Places (circles) which represent conditions. In this context they represent resource availability or process status.

In our example, places p_1 and p_2 indicate the availability status of the robot and parts respectively, and p_3 indicates the operational status of the robot.

Transitions(bars) which represent events. For our purposes, they represent the starting and stopping of processes.

t_1 in the example represents the start of the robot moving a part; t_2 the end of the robot moving the part.

Input functions that define arcs *(arrows)* from places to transitions.

The arc from p_1 to t_1 defines p_1 as an input place to transition t_1; similarly the arc from p_2 to t_1 defines p_2 as an input place to t_1. The arc from p_3 to t_2 defines p_3 as an input place to t_2.

Output functions that define arcs *(arrows)* from transitions to places.

The arc from t_1 to p_3 defines p_3 as an output place from t_1 and the arc from t_2 to p_1 defines p_1 as an output place from t_1.

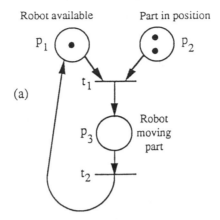

(a)

t_1: Robot starts moving part
t_2: Robot finishes moving part

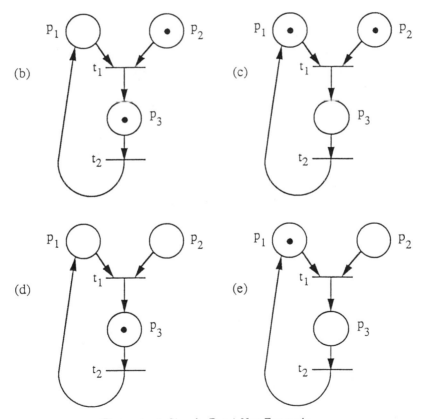

Figure 1. A Simple Petri Net Example.

These four elements define the structure of an ordinary Petri net. The state of a PN is indicated by it's:

> *Marking*, which is the number of *tokens (dots)* in each place. In general, there could be more than one token per place.

> The *initial* marking for the PN in Figure 1 is one token in place p_1, two tokens in place p_2, and zero tokens in p_3.

The state of the PN evolves from an *initial marking* according to the following execution rules:

> *Enabling:* If all the places that are inputs to a transition are marked then the transition is enabled and it may fire.

> *Firing:* When an enabled transition fires, a token is removed from each of the input places and a token is placed in each of the output places.

Thus far we have informally defined the *structure* of Petri nets in terms of places, transitions, and arcs and net *behavior* in terms of the marking and the rules for execution. Later these terms will be formally defined, with the appropriate mathematical notation. While the structure and behavior of Petri nets are defined, the interpretation, i.e., the meaning attached to the places and transitions, is left up to the modeler. In the example shown in Figure 1, the labels on the places and transitions give meaning to the Petri net.

The simple example of Figure 1 models a discrete control sequence for a robot to move two parts. Before it can start moving a part, the robot must be available and a part must be available. Once these two conditions are satisfied, the robot can move the part. Having finished moving the part, the robot is then available to move another part.

In Figure 1(a), places p_1 and p_2 are the input places to transition t_1. The token in p_1 indicates that the robot is available and the tokens in p_2 indicate that two parts are available. This is the initial marking for this Petri net. Note that this initial marking enables only transition t_1. It would not be enabled if only p_1 or only p_2 had tokens. Since t_1 is enabled, it can fire according to the rule for execution. When it does, one token from each of t_1's input places, p_1 and p_2, are removed and a token is placed in p_3, t_1's only output place.

Figure 1(b) shows the marked Petri net resulting from the firing of transition t_1. The token in place p_1 indicates that one part is available. The token in place p_3 in Figure 1(b) indicates the process of the robot moving one part. This marking enables only transition t_2. This transition is fired when the robot finishes moving the part. The firing of t_2 causes the token to be removed from p_3, t_2's only input place, and a token to be placed in t_2's only output place, p_1. This represents release of the robot.

The result is the marking as shown in Figure 1(c) with one token in p_1 and one token in p_2. Thus, t_1 is again enabled. When it fires, a token is removed from both p_1 and p_2 and a token is placed in p_3. This indicates the second part being moved. t_2 is again enabled and when it fires the token in p_3 is removed and a token is placed in p_1. Thus, the robot has moved the two parts. Note that in this state no transition is enabled. This is known as deadlock. In the context of this example, this is fine. However, in most manufacturing applications this is undesirable.

This example has illustrated some of the fundamentals of Petri nets. For each of the transitions the input places act as a set of preconditions. Once these are satisfied, i.e., the input places are marked, when and if the transition fires, the postconditions are set true, i.e., the output places are marked. This illustrates synchronization, two or more conditions must be true at the same time (concurrently) before the system or process may continue. It is easy to see how this mechanism can be used to model the precedence relations inherent in manufacturing processes. The example used places p_1 and p_2 to represent conditions and p_3 to represent process status.

II.   MODELING MANUFACTURING SYSTEMS

In manufacturing the process plan for each product is essential to achieve co-ordination. The process plan details all the activities, the resources required to perform them and the order, if important, in which they must occur. It may include specific machining operations and which machines can perform them, part routing, assembly sequencing, and so on. The process plans embody a complex set of precedence relations between activities and can give rise to a challenging resource allocation problem because of the high degree of concurrency and interaction.

A.   MODELING METHODOLOGY

Modeling methodology is key to the application of Petri nets to manufacturing systems. We start with one approach to modeling. Other approaches are possible.

1. Identify the activities and resources required for the production of one item of each product.

    In the simple example of the previous section, there is only one activity, *robot moving part*, and two resources, *a robot* and *parts*.

2. Order activities by the precedence relations as given in the process plans.

3. For each activity in order: create and label a place to represent the status of that activity; add a transition (start activity) with an output arc(s) to the place(s); add a transition (stop activity) with an input arc(s)

from the activity place(s). In general, the stop transition for one activity will be the same as the start transition for the next activity. Note that activity streams with no precedence relation between them will not be connected.

When the net is executed, a token in an activity place will indicate the activity is taking place. Multiple tokens will indicate the activity occurring in multiplicity, for example, in a machining activity place, two tokens might represent two parts being machined at the same time. The firing of the start transition represents starting the activity or process and the firing of the stop transition represents the completion of the activity and may also represent the start of the next activity.

For the example, create and label a place p_3 to represent *robot moving part*. Add transition t_1 with an output arc to p_1 and add transition t_2 with an input arc from p_3.

4. For each activity in order: If such a place has not been already created, create and label a place, for each resource which must be available to start the activity. Connect all appropriate resource availability places with arcs such that each inputs to the starting transition for the activity. Create output arcs to connect the stop transition following the activity to any resource places representing resources which become available (are released) upon completion of the activity.

   For the example, the resources required for the activity *robot moving part* are available in position and the robot available. Places p_1 and p_2 are created to model these, respectively.

5. Specify the initial marking for the system.

   For the example the initial conditions are: the robot is available and two parts are in position. This requires a single token in p_1 and two tokens in p_2.

B. A MODELING EXAMPLE

To illustrate the modeling methodology in a more general way, take a simple manufacturing system. It consists of two machines, M_1 and M_2, and a shared robot, R, for loading and unloading the machines. The system produces two parts, A and B. The process plan shows A is to be machined on M_1 and B is to be machined on M_2 with no precedence relation between them. Following the methodology:

1. The activities required are loading, machining, and unloading and the resources are M_1, M_2, R and raw stock for A and B.

2. The order of activities is as follows:

Part A
A1: Load raw stock A on M_1
A2: Process A on M_1
A3: Unload A from M_1

Part B
B1: Load raw stock B on M_2
B2: Process B on M_2
B3: Unload B from M_2

For each part the activities are ordered as listed and the two parts' activities could occur concurrently if resources are available.

3. As shown in Figure 2 and Table 1, places p_1, p_2, and p_3 are created to model the activity sequence for one part of type A. Transition t_1 models the start of load A activity; t_2 models the stop of load A and the start of process A on M_1, t_3, the stop of process A and the start of unload A from M_1, and t_4, the stop of unload A from M_1. Places p_4, p_5, p_6 and transitions t_5, t_6, t_7, and t_8 model similar activities and starts and stops for the activity sequence for one part of type B. Note that these two activity sequences are not connected. This indicates the possible concurrency between them.

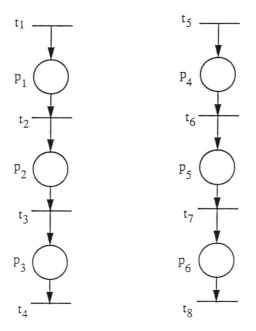

Figure 2. Petri Net Graph of the
Activity Sequences for the Manufacturing Example.

4. To load raw stock A onto M_1 we require raw stock A, the robot, R, and the machine,

p_1	Load raw stock A on M_1
p_2	Process A on M_1
p_3	Unload A from M_1
p_4	Load raw stock B on M_2
p_5	Process B on M_2
p_6	Unload B from M_2
p_7	Raw stock A available
p_8	R available
p_9	M_1 available
p_{10}	Raw stock B available
p_{11}	M_2 available

Table 1. Place labels for Figures 2 and 3.

M_1, all be available. Figure 3 shows the Petri net model with places p_7, p_8, and p_9 representing these resource availabilities, respectively. Each has an input arc to transition t_1. Next consider t_2 as the stop transition for activity A1. When the loading activity is complete, the robot can be released and becomes available again. This is modeled as the output arc from t_2 to p_8. Now consider t_2 as the start transition of A2. No further resources are required for A2 so no other input arcs are added.

For t_3 no resources are released when activity A2 stops. Therefore, no output arcs are added. To start activity A3, R must be available. This is modeled by an input arc from p_8 to t_3. When A_3 stops, both R and M_1 are released and become available again. This is modeled by output arcs from t_4 to p_8 and t_4 to p_9, respectively.

Cyclical behavior can be modeled by the assumption of an infinite supply of raw stock. An output arc from t_4 to p_7 models this assumption.

The reader can continue the procedure to model part B activities and create p_{10}, p_{11}, and the remaining arcs shown in Figure 3.

5. The initial marking is formulated to start the system. The raw materials, the machines, and the robot should be available to start. The initial marking for these conditions is shown in Figure 3. Note that in this marking transitions t_1 and t_5 are both enabled.

C.   EXECUTING THE MODEL

The dynamic behavior of the net system can be observed by executing the net by using the rules and generating possible activity sequences for the system.

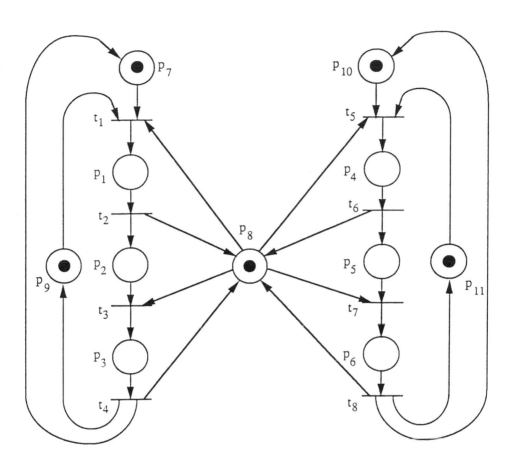

Figure 3. Petri Net Graphs.

To start, both t_1 and t_5 are enabled. However, if we fire one of these, the other will be disabled because the initial marking has only one token in p_8. This models the situation that only one machine at a time can be loaded or unloaded by the single robot. The robot is a shared resource. The Petri net structure guarantees a mutual exclusion, that is when one activity is allocated to the robot all other activities are excluded from using it. Place p_8 is called a *choice* place (also known as a conflict or decision place). Any place with more than one arc out of it is a choice place. Note that in the case of p_8 the net cannot make this choice. The *token player* or *executor* must make this decision. If this net were executed on a computer, then conflict resolution software or user interaction must be added to the token player.

In this instance the authors will act as token players. t_1 is fired first. Following the execution rules given earlier, one token is removed from each of p_7, p_8, and p_9 and a token is put in p_1. The new marking indicates that M_1 is being loaded with raw stock A by R. In this state only t_2 is enabled. This information and that for the remaining firings by the token player are summarized in Table 2. t_2 is fired next. The new marking shows that M_1 is processing part A and that both t_3 and t_5 are enabled. t_5 is chosen and fired. The resulting marking indicates that M_1 is processing part A and raw stock B is being loaded by R. Note that this explicitly models concurrent activities, a powerful advantage of Petri nets over finite state machine representations. In this state t_6 is enabled. t_6 is chosen and fired. See Table 2 to continue.

The reader is encouraged to play the token game for this net and others using popcorn or split peas to move the tokens through the net.

Transition fired:	Remove token from places:	Add token to places:	Transitions enabled by new marking:
t_1(choice)	p_7, p_8, p_9	p_1	t_2
t_2	p_1	p_2, p_8	t_3, t_5
t_5(choice)	p_8, p_{10}, p_{11}	p_4	t_6
t_6	p_4	p_8, p_5	t_3, t_7
t_3(choice)	p_2, p_8	p_3	t_4
t_4	p_3	p_7, p_8, p_9	t_7, t_1
.	.	.	.
.	.	.	.

Table 2. A Sample Sequence from the Petri Net Example.

D.   RELEVANT PROPERTIES OF PETRI NETS

One of the major advantages of Petri nets is the ability to analyze these nets for properties related to manufacturing control. These properties include reachability, boundedness, liveness and reversibility as defined and discussed as follows. These definitions follow those of Murata [1]. This reference is an excellent tutorial on Petri nets.

Reachability: A marking M_d is said to be reachable from a marking M_o if there exists a sequence of firings that transforms M_o to M_d. A firing sequence is denoted by σ and is a list of transitions in order of firing from M_o. M_d is said to be reachable from M_o by σ. $R(M_o)$ is the set of all possible reachable markings (reachability set) for a particular net, N.

Boundedness: A Petri net, N, with initial marking M_o is *bounded* if the number of tokens in each place does not exceed a finite number k for any marking reachable from M_o, i.e., $M_o(p) \leq k \ \forall \ p \ \epsilon \ P$ and $\forall \ M \ \epsilon \ R(M_o)$. A Petri net, N, is safe if it is 1-bounded.

In manufacturing, boundedness implies that there will be no overflows. For example, no buffers or machines will exceed their capacity to handle parts. Note that the net shown in Fig. 1 is bounded.

Liveness: A Petri net, N, is *live* if no matter what marking has been reached from M_o, it is possible to fire ultimately any transition of the net by progressing through some further firing sequence. A live Petri net guarantees deadlock free operation, no matter what firing sequence is chosen.

The application to manufacturing is direct because, in general, the discrete event control should not lead to deadlock or at least all possibilities for deadlock should be known.

III.   ANALYSIS OF PETRI NETS

A.   PETRI NET DEFINITIONS

A Petri net is a 5 tuple:

$$N = \{P, T, I, O, M_0\} \text{ where:}$$

$P = \{p_1, p_2, \ldots, p_m\}$ is a finite set of places,

$T - \{t_1, t_2, \ldots, t_n\}$ is a finite set of transitions,

$P \cup T \neq \phi, \ P \cap T = \phi$,

$I : P \times T \rightarrow \{0, 1, 2, \ldots\}$ is an input function that defines directed arcs from places to transitions, i.e., if $I(p_i, t_j) > 0 = k$, then we include an arc from p_i to t_i. If $I(p_i, t_j) > 1$, then we label the arc k.

$O : P \times T \rightarrow \{0, 1, 2, \ldots\}$ is an output function that defines directed arcs from transitions to places, i.e.,

if $O(p_i, t_j) > 0 = k$, then we include an arc from t_i to p_i. If $O(p_i, t_j) > 1$, then we label the arc k.

$M_0 : P \rightarrow \{0, 1, 2, \ldots\}$ is the initial marking where the marking M indicates the number of tokens in each place.

Note: The class of nets where we allow $I(p_i, t_j) > 1$ are known as generalized Petri nets. When $I(p, t) = 0$ or 1, the class is known as ordinary Petri nets.

Consider the example given in Figure 4. This is the same as the example in Figure 1 except it includes the assumption of infinite supply of parts.

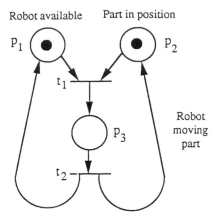

Figure 4. Simple Example with Assumption of Infinite Supply of Parts.

$$P = \{p_1, p_2, p_3\}$$

$$T = \{t_1, t_2\}$$

$I(p_1, t_1) = I(p_2, t_1) = I_3(p_3, t_2) = 1$

$I(p_3, t_1) = I(p_1, t_2) = I(p_2, t_2) = 0$

This is an ordinary PN since the arc weights are 1.

Matrix form:

$$I(p,t) = \begin{array}{c} \\ p_1 \\ p_2 \\ p_3 \end{array} \begin{array}{cc} t_1 & t_2 \\ \left[\begin{array}{cc} 1 & 0 \\ 1 & 0 \\ 0 & 1 \end{array} \right] \end{array}$$

Similarly,

$$0(p_1, t_2) = 0(p_2, t_2) = 0(p_3, t_1) = 1$$

$$0(p_3, t_2) = 0(p_1, t_1) = 0(p_2, t_1) = 0$$

$$0(p,t) = \left[\begin{array}{cc} 0 & 1 \\ 0 & 1 \\ 1 & 0 \end{array} \right]$$

$$M_0 = \begin{matrix} p_1 \\ p_2 \\ p_3 \end{matrix} \begin{bmatrix} 1 \\ 1 \\ 0 \end{bmatrix}$$

Enabling rule: A transition, t_j, of a PN is said to be enabled in a marking M if and only if

$M(p_i) \geq I(p_i, t_j)$ for all p_i which are members of the set of input places of t_j.

Firing rule: An enabled transition can fire at any time (we control the firing). When a transition, t_j, enabled in a marking M fires, a new marking M' is reached according to the equation:

$$M'(p_1) = M(p_i) + 0(p_i, t_j) - I(p_i, t_j) \; \forall p_i \epsilon P$$

We say M' is reachable from M.
For the example in Figure 4:
Using the enabling rule from M_0:

$$M_0 = \begin{bmatrix} 1 \\ 1 \\ 0 \end{bmatrix} \geq \begin{bmatrix} 1 \\ 1 \\ 0 \end{bmatrix} = I(p, t_1), \; \forall \; p \; \epsilon \; P,$$

therefore t_1 is enabled;

$$M_0 = \begin{bmatrix} 1 \\ 1 \\ 0 \end{bmatrix} \ngeq \begin{bmatrix} 0 \\ 0 \\ 1 \end{bmatrix} = I(p, t_2), \; \forall \; p \; \epsilon \; P,$$

therefore t_2 is not enabled.
When we fire t_1 from M_0 using the firing rule:

$$M_1 = M_0 + 0(p, t_1) - I(p, t_1)$$

$$M_1 = \begin{bmatrix} 1 \\ 1 \\ 0 \end{bmatrix} + \begin{bmatrix} 0 \\ 0 \\ 1 \end{bmatrix} - \begin{bmatrix} 1 \\ 1 \\ 0 \end{bmatrix} = \begin{bmatrix} 0 \\ 0 \\ 1 \end{bmatrix}$$

Now from M_1:

$$M_1 = \begin{bmatrix} 0 \\ 0 \\ 1 \end{bmatrix} \geq \begin{bmatrix} 0 \\ 0 \\ 1 \end{bmatrix} = I(p, t_2), \; \forall \; p \; \epsilon \; P,$$

therefore t_2 is enabled;

$$M_1 = \begin{bmatrix} 0 \\ 0 \\ 1 \end{bmatrix} \not\geq \begin{bmatrix} 1 \\ 1 \\ 0 \end{bmatrix} = I(p, t_1), \; \forall \; p \; \epsilon \; P,$$

therefore t_2 is not enabled.
If we fire t_2 from M_1:

$$M_2 = \begin{bmatrix} 0 \\ 0 \\ 1 \end{bmatrix} + \begin{bmatrix} 1 \\ 1 \\ 0 \end{bmatrix} - \begin{bmatrix} 0 \\ 0 \\ 1 \end{bmatrix} = \begin{bmatrix} 1 \\ 1 \\ 0 \end{bmatrix},$$

and so on.

B.   THE MATRIX FORM OF PETRI NETS

Incidence Matrix: For a PN N with m places and n transitions we can define an incidence matrix:

$$C = C(p_i, t_j) = O(p_i, t_j) - I(p_i, t_j)$$

For our example:

$$C(p_i, t_j) = \begin{bmatrix} 0 & 1 \\ 0 & 1 \\ 1 & 0 \end{bmatrix} - \begin{bmatrix} 1 & 0 \\ 1 & 0 \\ 0 & 1 \end{bmatrix} = \begin{bmatrix} -1 & 1 \\ -1 & 1 \\ 1 & -1 \end{bmatrix}$$

We can define u_k as the kth firing or control vector, all of whose elements are zero except the ith element which is one, to indicate transition i fires at the kth firing, then

$$M_k = M_{k-1} + C u_k.$$

For our example:

$$M_1 = M_0 + C u_1, \quad \text{firing } t_1 :$$

$$M_1 = \begin{bmatrix} 1 \\ 1 \\ 0 \end{bmatrix} + \begin{bmatrix} -1 & 1 \\ -1 & 1 \\ 1 & -1 \end{bmatrix} \begin{bmatrix} 1 \\ 0 \end{bmatrix} = \begin{bmatrix} 1 \\ 1 \\ 0 \end{bmatrix} + \begin{bmatrix} -1 \\ -1 \\ 1 \end{bmatrix} = \begin{bmatrix} 0 \\ 0 \\ 1 \end{bmatrix}$$

Now fire t_2, the only transition enabled at M_1:

$$M_2 = \begin{bmatrix} 0 \\ 0 \\ 1 \end{bmatrix} + \begin{bmatrix} -1 & 1 \\ -1 & 1 \\ 1 & -1 \end{bmatrix} \begin{bmatrix} 0 \\ 1 \end{bmatrix} = \begin{bmatrix} 0 \\ 0 \\ 1 \end{bmatrix} + \begin{bmatrix} 1 \\ 1 \\ -1 \end{bmatrix} = \begin{bmatrix} 1 \\ 1 \\ 0 \end{bmatrix}$$

and so on.

We can generalize this in symbolic form:

$$
\begin{aligned}
M_1 &= M_o + Cu_1, \\
M_2 &= M_1 + Cu_2 = M_o + Cu_1 + Cu_2 \\
&= M_o + C(u_1 + u_2), \\
M_3 &= M_2 + Cu_3 = M_o + C(u_1 + u_2) + Cu_3 \\
&= M_o + C\sum_{k=1}^{3} u_k
\end{aligned}
$$

or after d firings, $u_1, u_2, \ldots, u_d$ we reach:

$$M_d = M_0 + Cf_d$$

where

$$\sum_{k=1}^{d} u_k = f_d \text{ is the firing count vector.}$$

Note that in this form we only know how many times each transition has been fired and not necessarily the sequence.

This form will be used later for analysis.

C. REACHABILITY ANALYSIS

The reachability tree: represents the reachability set (all reachable markings from M_o) of a Petri net.

As an example, the reachability tree for the net in Figure 4 is given in Figure 5.

Note that the reachability tree is infinite.

To illustrate construction of the reachability tree, we take a net given in Petersen [2] (Figure 6). This book gives an excellent introduction to Petri net theory.

We start at the initial marking and fire each enabled transition from that marking and graph the results as shown in Figure 7.

From each of the markings reached in the previous step, each enabled transition is fired and the new marking graphed. The next two steps are shown in Figure 8.

If we complete the reachability tree, it provides detailed information about the net behavior. It gives:

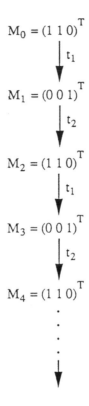

$$M_0 = (1\ 1\ 0)^T$$

$$\downarrow t_1$$

$$M_1 = (0\ 0\ 1)^T$$

$$\downarrow t_2$$

$$M_2 = (1\ 1\ 0)^T$$

$$\downarrow t_1$$

$$M_3 = (0\ 0\ 1)^T$$

$$\downarrow t_2$$

$$M_4 = (1\ 1\ 0)^T$$

Figure 5. Reachability Tree of Net in Figure 4.

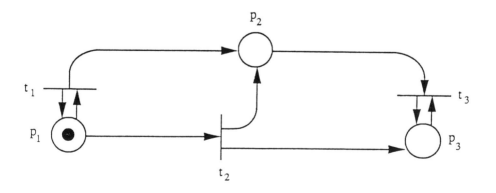

Figure 6. Net to Illustrate Reachability Tree.

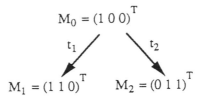

Figure 7. First Step in Construction.

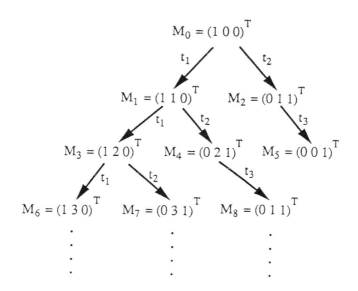

Figure 8. First Three Steps of Construction of Reachability
Tree of Net in Figure 6.

1. All possible firing sequences, for the example in Figure 8:

$$
\begin{array}{llll}
t_1 & t_1 & t_1 & \ldots \\
t_1 & t_1 & t_2 & \ldots \\
t_1 & t_2 & t_3 & \ldots \\
t_2 & t_3 &
\end{array}
$$

In this example not all sequences are given because the net is incomplete and there may be an infinity of sequences.

2. All reachable states in each sequence. For the example:

$$
\begin{array}{llll}
M_o & M_1 & M_3 & M_6 & \ldots \\
M_o & M_1 & M_3 & M_7 & \ldots \\
M_o & M_2 & M_4 & M_8 & \ldots \\
M_o & M_2 & M_5 &
\end{array}
$$

3. The combination of the above

   For the example:

$$
M_o t_1 \; M_1 t_1 \; M_3 t_1 \; M_6 \; \ldots
$$

the reader can fill in the remaining sequence information.

As the preceding shows, the reachability tree is straightforward to construct but the method given may result in an infinite tree. Even a tree with a finite reachability set can be infinite, e.g., the tree in Figure 5.

We need a finite representation of reachability information. Certain classes of nodes will limit the new markings produced at each step.

1. Dead-ends (terminal nodes). Ex. M_5 in Figure 8.

2. Duplicate nodes or nodes which have previously appeared. Ex. M_8 is a duplicate of M_2 in Figure 8. Since all successors will be produced by the first occurrence, M_2, no successors to the duplicate need be considered.

3. Infinite loops. Consider a sequence of transition firings which starts at M and ends at M' where $M' > M$ or M' covers M.

   Ex. In Figure 8

$$
M_6 > M_3 > M_1 > M_0
$$
$$
\begin{bmatrix} 1 \\ 3 \\ 0 \end{bmatrix} > \begin{bmatrix} 1 \\ 2 \\ 0 \end{bmatrix} > \begin{bmatrix} 1 \\ 1 \\ 0 \end{bmatrix} > \begin{bmatrix} 1 \\ 0 \\ 0 \end{bmatrix}
$$

The marking M' is the same as M except it has "extra" in some places. Since transition firing is not affected by extra tokens, the sequence σ can be fired again and again producing an arbitrarily large number of tokens. These are infinite loops. We represent the infinite number of tokens by using a special symbol ω which can be through of as "infinity". It is defined for any constant a

$$
\begin{aligned}
w + a &= \omega \\
w - a &= \omega \\
a &< \omega \\
\omega &\leq \omega.
\end{aligned}
$$

With the symbol ω, a coverability tree algorithm for a finite representation can be developed as found in Peterson [2] and in Murata [1].

Step 1 Label the initial marking M_0 as the root and tag it "new".

Step 2 While "new" markings exist, do the following:

    Step 2.1 Select a new marking M.

    Step 2.2 If M is identical to a marking on the path from the root to M, then tag M "old" and go to another new marking.

    Step 2.3 If no transitions are enabled at M, tag M "deadend".

    Step 2.4 While there exist enabled transitions at M, do the following for each enabled transition t at M:

        Step 2.4.1 Obtain the marking M' that results from firing t at M.

        Step 2.4.2 On the path from the root to M if there exists a marking M'' such that $M'(p) \geq M''(p)$ for each place p and $M' \neq M''$, i.e., M' is coverable, then replace $M'(p)$ by ω for each p such that $M'(p) > M''(p)$.

        Step 2.4.3 Introduce M' as a node, draw an arc with label t from M to M', and tag M' "new".

We can take the net given in Figure 6 and construct the coverability tree as given in Figure 9.

This is now a finite representation of what would have been an infinite reachability tree. Note that if the net is unbounded as in Figure 6 then we lose information as in Figure 9 because all sequences and reachable markings are not shown. This is not important in manufacturing because in general, if the net is unbounded, we have made a mistake in modeling or system design.

Some properties that can be checked using the coverability tree are as follows:

1. A net, N, is bounded and thus $R(M_0)$ is finite if and only if ω does not appear in any node labels in the tree.

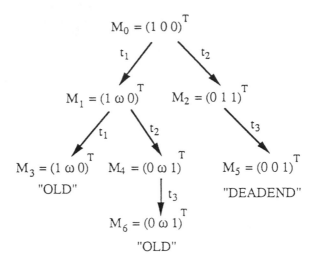

Figure 9. Coverability Tree Example.

2. A net, N, is safe iff only 0's and 1's appear in node labels in the tree.

3. A transition t is dead iff it does not appear as an arc label in the tree.

4. If M is reachable from M_0, then there exists a node labeled M' such that $M \leq M'$.

With this method it is easy to check boundedness, liveness and reversibility of a Petri net.

Take the manufacturing system example whose net is shown in Figure 3. The coverability tree is shown in Figure 10. Since no ω's appear in the net we know the net is bounded. In fact since all markings in the reachability set contain only 0's and 1's, we know that the net is safe. Since there are no deadend states we have no possibility for deadlock, i.e., the system is live. Finally, the net is reversible because from any marking we can find a firing sequence that brings the system to M_0, the initial marking.

The disadvantage of the reachability tree is that it enumerates all possible markings and for a complex system, this may be computationally impossible or impractical. However, some things may be shown about the net using matrix analysis.

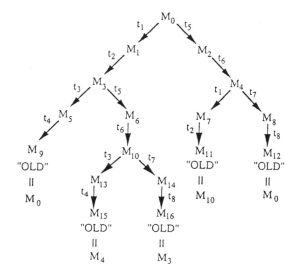

$M_0 = [0\ 0\ 0\ 0\ 0\ 0\ 1\ 1\ 1\ 1\ 1]^T$

$M_1 = [1\ 0\ 0\ 0\ 0\ 0\ 0\ 0\ 0\ 1\ 1]^T$

$M_2 = [0\ 0\ 0\ 1\ 0\ 0\ 1\ 0\ 1\ 0\ 0]^T$

$M_3 = [0\ 1\ 0\ 0\ 0\ 0\ 0\ 1\ 0\ 1\ 1]^T$

$M_4 = [0\ 0\ 0\ 0\ 1\ 0\ 1\ 1\ 1\ 0\ 0]^T$

$M_5 = [0\ 0\ 1\ 0\ 0\ 0\ 0\ 0\ 0\ 1\ 1]^T$

$M_6 = [0\ 1\ 0\ 1\ 0\ 0\ 0\ 0\ 0\ 0\ 0]^T$

$M_7 = [1\ 0\ 0\ 0\ 1\ 0\ 0\ 0\ 0\ 0\ 0]^T$

$M_8 = [0\ 0\ 0\ 0\ 0\ 1\ 1\ 0\ 1\ 0\ 0]^T$

$M_9 = M_0$

$M_{10} = [0\ 1\ 0\ 0\ 1\ 0\ 0\ 1\ 0\ 0\ 0]^T$

$M_{11} = M_{10}$

$M_{12} = M_0$

$M_{13} = [0\ 0\ 1\ 0\ 1\ 0\ 0\ 0\ 0\ 0\ 0]^T$

$M_{14} = [0\ 1\ 0\ 0\ 0\ 1\ 0\ 0\ 0\ 0\ 0]^T$

$M_{15} = M_4$

$M_{16} = M_3$

Figure 10. Coverability Tree for the Petri Net Shown in Figure 3

D. MATRIX ANALYSIS

Matrix analysis sometimes called invariant analysis may allow comments on boundedness, liveness, and reversibility of the system model.

Take the matrix form for the Petri net model as shown earlier. For any marking M

$$M = M_0 + Cf$$

Suppose we premultiply this equation by x^T.

$$x^T M = x^T M_0 + x^T C f$$

We notice that if

$$x^T C = 0$$

then $x^T M = x^T M_0$.

What this says is that if we can find values of the x vector such that the homogeneous equation is satisfied then the weighted sum of the markings will be constant or invariant over all markings in the reachability set. Invariants that satisfy the above conditions are called place invariants since for any invariant the weighted sum of the tokens in its places are constant. The reader is referred to Murata [1], Lautenbach [3] for more complete discussions of invariants and to Narahari and Viswanadham [4] for examples of their application to manufacturing.

We can show boundedness of a net using place (P) invariants. Informally, since the weighted sum of the tokens in the places of a P invariant is constant for all markings, then all of these places are bounded. Then if all places in the net are covered by at least one invariant, the net is bounded.

Take the manufacturing example given earlier. The incidence matrix is:

$$
C = \begin{bmatrix}
-1 & 1 & 0 & 0 & 0 & 0 & 0 & 0 \\
0 & -1 & 1 & 0 & 0 & 0 & 0 & 0 \\
0 & 0 & -1 & 1 & 0 & 0 & 0 & 0 \\
0 & 0 & 0 & 0 & -1 & 1 & 0 & 0 \\
0 & 0 & 0 & 0 & 0 & -1 & 1 & 0 \\
0 & 0 & 0 & 0 & 0 & 0 & -1 & 1 \\
1 & 0 & 0 & -1 & 0 & 0 & 0 & 0 \\
1 & -1 & 1 & -1 & 1 & -1 & 1 & -1 \\
1 & 0 & 0 & -1 & 0 & 0 & 0 & 0 \\
0 & 0 & 0 & 0 & 1 & 0 & 0 & -1 \\
0 & 0 & 0 & 0 & 1 & 0 & 0 & -1
\end{bmatrix}
$$

Then let $x = (x_1\ x_2\ x_3\ x_4\ x_5\ x_6\ x_7\ x_8\ x_9\ x_{10}\ x_{11})^T$ be a P-invariant such that $x^T C = 0$, or

$$-x_1 + x_7 + x_8 + x_9 = 0$$

$$\left.\begin{aligned} x_1 - x_2 - x_8 &= 0 \\ x_2 - x_3 + x_8 &= 0 \\ x_3 - x_7 - x_8 - x_9 &= 0 \end{aligned}\right\} \quad x_1 - x_7 - x_8 - x_9 = 0$$

$$-x_4 + x_8 + x_{10} + x_{11} = 0$$

$$\left.\begin{aligned} x_4 - x_5 - x_8 &= 0 \\ x_5 - x_6 + x_8 &= 0 \\ x_6 - x_8 - x_{10} - x_{11} &= 0 \end{aligned}\right\} \quad x_4 - x_8 - x_{10} - x_{11} = 0$$

Solution to the preceding equations yields the minimal P-invariants:

$$\begin{aligned} I_1 &= (\ 1\ \ 1\ \ 1\ \ 0\ \ 0\ \ 0\ \ 1\ \ 0\ \ 0\ \ 0\ \ 0\ )^T \\ I_2 &= (\ 1\ \ 1\ \ 1\ \ 0\ \ 0\ \ 0\ \ 0\ \ 0\ \ 1\ \ 0\ \ 0\ )^T \\ I_3 &= (\ 1\ \ 0\ \ 1\ \ 1\ \ 0\ \ 1\ \ 0\ \ 1\ \ 0\ \ 0\ \ 0\ )^T \\ I_4 &= (\ 0\ \ 0\ \ 0\ \ 1\ \ 1\ \ 1\ \ 0\ \ 0\ \ 0\ \ 1\ \ 0\ )^T \\ I_5 &= (\ 0\ \ 0\ \ 0\ \ 1\ \ 1\ \ 1\ \ 0\ \ 0\ \ 0\ \ 0\ \ 1\ )^T \end{aligned}$$

To interpret the invariants substitute each of these into

$$x^T M = x^T M_0$$

which was given earlier
For I_1:

$$M(p_1) + M(p_2) + M(p_3) + M(p_7) = 1$$

This says that the sum of the tokens in places 1, 2, 3, and 7 equals one for any marking. This means that only one of these places can have a token in any given marking and therefore each of these places is safe. Physically this means that either Part A raw stock is waiting (p_7) or it is being loaded (p_1) or Part A is being processed (p_2) or Part A is being unloaded (p_3). Similarly for I_2, I_3, I_4, and I_5:

$$M(p_1) + M(p_2) + M(p_3) + M(p_9) = 1$$

$$M(p_1) + M(p_3) + M(p_4) + M(p_6) + M(p_8) = 1$$

$$M(p_4) + M(p_5) + M(p_6) + M(p_{10}) = 1$$

$$M(p_4) + M(p_5) + M(p_6) + M(p_{11}) = 1$$

The reader is encouraged to interpret each of these equations as to their physical significance. It is important to note that all places are covered by an invariant and hence the net is bounded. In fact, the net is safe since all sums are equal to one. P-invariants can also be used as constraints in net liveness proofs and an excellent example is given in Narahari and Vishwanadham [4].

Again taking the matrix form:

$$M = M_0 + Cf$$

if we can solve this for firing vectors y such that

$$C y = 0$$

then those values of y for which this is true are called T or transition invariants. For such values of y:

$$M = M_0.$$

The existence of T invariants covering all transitions of the net is necessary but *not* sufficient to show reversibility. The reason is that we may find firing count vector solutions (T invariants) that are not firable.

For our manufacturing example,

Let $y = (y_1\ y_2\ y_3\ y_4\ y_5\ y_6\ y_7\ y_8)^T$ to be a T-invariant such that $Cy = 0$, then

$$\left.\begin{array}{rcl} -y_1 + y_2 & = & 0 \\ -y_2 + y_3 & = & 0 \\ -y_3 + y_4 & = & 0 \end{array}\right\} y_1 = y_2 = y_3 = y_4$$

$$\left.\begin{array}{rcl} -y_5 + y_6 & = & 0 \\ -y_6 + y_7 & = & 0 \\ -y_7 + y_8 & = & 0 \end{array}\right\} y_5 = y_6 = y_7 = y_8$$

This yields the following minimal T-invariants,

$$\begin{array}{rcl} TI_1 & = & (\ 1\ \ 1\ \ 1\ \ 1\ \ 0\ \ 0\ \ 0\ \ 0\ )^T \\ TI_2 & = & (\ 0\ \ 0\ \ 0\ \ 0\ \ 1\ \ 1\ \ 1\ \ 1\ )^T \end{array}$$

The first T-invariant indicates that t_1, t_2, t_3 and t_4 each fire once. We note from the net that there is a fireable sequence for this invariant and hence the net is reversible. The reader can look at the second T-invariant and interpret. Remember, if no T-invariant exists then we know the net is not reversible. Existence of T-invariants covering all transitions is not sufficient to show reversibility. We must also show that there exists fireable sequences which may be non-trivial.

E. REDUCTION AND SYNTHESIS

The reduction of the size of the net while maintaining properties such as bound-edness, liveness, and reversibility is a powerful way to reduce the complexity of the net for analysis purposes. The reader is referred to Murata [1], Silva [5], and Berthelot [6] for discussions of reduction.

Conversely, one method of design is to synthesize the nets and initial mark-ings in such a manner so as to guarantee these properties without the need for analysis. For a review of the past methods see Jeng [7] and for more advanced synthesis concepts relating to manufacturing see Zhou [8].

IV. CONTROL USING PETRI NETS

If a net can be coded into data and the computer has a token player, then Petri nets can be used to control physical systems with the addition of soft-ware drivers and wiring interfaces between the computer and the machines. The description given here comes from an FMS control implementation [9] at Rensselaer Polytechnic Institute. Other examples and discussion of implemen-tation can be found in [10, 11, 12, 13].

The method used here is to code the net using a Petri net description language and to supervise the system by employing a Petri net token player which uses the net description as data.

A. PETRI NET DESCRIPTION LANGUAGE (PNDL)

The Petri net description language is a declarative language which is developed and used to specify a Petri net. Using such a language, the graphical nota-tion of a Petri net can be converted to a text representation which eventually triggers the routines in the system. Each PNDL text file defines information about places, transitions, and the connectivity of the Petri net model. Two major statements are declaration statements and assignment statements. All places and transitions must be declared in declaration statements before their attributes are assigned in assignment statements.

A declaration specifies a type, and is followed by a list of one or more identifiers of that type. The identifiers in PNDL are the names of places and transitions. The type is given by one of the keywords: *pl-simple*, *pl-action*, and *trans*. *pl-simple* declares places to represent resources or buffers and *pl-action* declares them to represent operations. All transitions are of type *trans*.

Places and transitions are given their attributes, which define information about a Petri net, by using assignment statements. An assignment statement consists of an identifier followed by a group of attribute statements in braces. Each attribute statement consists of an attribute type followed by a list of one or more identifiers or constants. The type is given by one of the attribute keywords. Examples of these keywords are as follows:

descr: literal string uniquely describing the place or the transition;

inp: identifier list of input places of a transition;

outp: identifier list of output places to a transition; and

mark: integer number of initial tokens marking a place.

For example, $T1$ (inp p_1, p_2; outp p_4; descr "Finish assembling") means that transition $T1$ has input places p_1 and p_2, output place p_4, and means "finishing assembling". These descriptions are processed into place and transition files. The reader is referred to [9] for further detail.

B. PETRI NET SUPERVISORY SYSTEM (PNSS)

A Petri net supervisory system is designed to execute the basic tasks of a system supervisor: to initiate and coordinate operations at workstations to perform a desired task. PNSS interprets the information contained in a Petri net model file and communicates with workstation controllers to initiate appropriate operations through the workstation computers. It consists of a Petri net execution algorithm and a set of PNSS commands. These commands are invoked to interpret a Petri net model described in a PNDL text file. They can also be used to interactively execute a Petri net model and to display information about the marking of the net. The kernel of the system is a token player program, i.e., the Petri net execution algorithm, which executes any net described by the data files. Written in the C programming language, PNSS runs on an IBM PC-XT computer and is used to successfully control the concurrent operations of the described FMS. The execution algorithm is explained as follows and its flow diagram is depicted in Figure 11.

The initialization includes setting the newly marked list of places (L) as the initially marked places, and the list of enabled transitions (E) empty. The elements in L will be ordered as a first-in-first-out queue. As shown in Figure 11, the first place in L is removed from L. If it is a resource place, its output transitions are checked to see if they are enabled. If none of its output transitions is enabled, it will be placed at the end of L. If any output transition is enabled, it is added to E. If there is any enabled transition, i.e., E is not empty, then one is selected to fire. Firing of a transition consists of unmarking its input places and removing them from L, marking its output places and adding them to L. If the place is an operation one, it is set to Phase-Start. Finally, all workstation computers are checked and a new iteration is started.

If the place type is *operation* as opposed to *resource*, it is checked to see if it is set to Phase-Start. If the place is in Phase-Start, the program will request an operation at the appropriate workstation, and will then poll all workstations for completed operations. Each place set to Phase-Start will remain set until a completion signal is returned by its workstation computer. In each iteration all operations are checked for completion and the places representing operations which are completed are set to Phase-Completion and are added to the beginning of L. When $L = \phi$, the algorithm checks the workstation computers

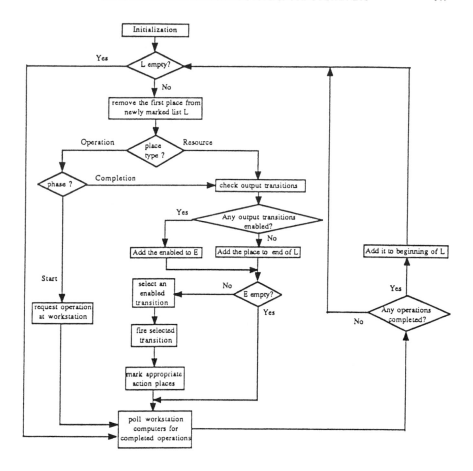

Figure 11. The Petri Net Execution Algorithm

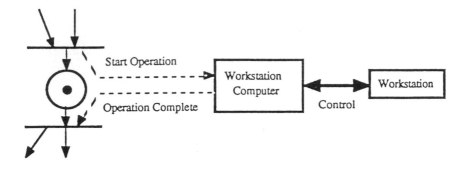

Figure 12. Interface Between the PN Supervisor
and a Workstation Computer

repeatedly. The change in status from Phase-Start to Phase-Completion allows
a marked place to enable its output transitions. If the operation place is set
to Phase-Completion, the algorithm then checks its output transitions and so
on, as in the case of a resource place as discussed above.

For the case of more than one transition enabled at a time, there can be
any number of selection mechanisms for which to fire, for example, a First-
In-First-Out (FIFO) policy and a priority policy. The details of implemented
selection mechanisms and software documentation can be found in [9].

As shown conceptually in Figure 12. when an operation place is marked,
the place is marked and a start is sent to the workstation computer. The
place is in Phase-Start until a completion signal is returned by the workstation computer. The above methods were used to control a moderately sized
FMS[14]. The Petri net model had about 40 transitions and 70 places. Control
was implemented using a host computer, an IBM PC-XT, called the system
supervisor in this FMS. This sent messages to six lower level control computers which were interfaced to the machines, robots and materials movement and
storage systems.

V. MARKOV MODELS

Models for manufacturing systems should be able to represent finite buffer
capacity, random machine failures and repairs, and should be useful for calculating various performance measures. Typical measures include throughput,
average machine utilization, probability that a machine is blocked or starved,
and expected time to complete a job. This section concentrates on the Markov
chain approach for both the discrete-transition case and the continuous-time

case. Although the Markov chain method requires the tedious enumeration of all possible states, it will be shown later how Petri nets can be used to automatically generate all the states in the chain. More detail on Markov models can be found in [15-19].

A.   SYSTEM STATE

A Markov model consists of a set of discrete states. This set is exhaustive and describes all the possible states in which the system can be. Transitions from state i to j occur with probability p_{ij}.

Example: A Simple Transfer Line

The state of a manufacturing system can be defined by considering a simple transfer line that consists of two machines (M_1 and M_2) and one buffer as shown in Figure 13.

It is assumed that the first machine is never starved, the last is never blocked, and the buffer has a finite capacity.

Each machine is described by a processing time, a time to fail, and a time to repair. If these quantities are deterministic, we have a discrete-state discrete-transition Markov chain (or a discrete-time Markov chain). They could also be random variables that would lead to a continuous-time Markov chain.

We can demonstrate the basic idea of a discrete-state discrete-transition Markov chain by modeling just one of the machines. In this case, the state of a machine is described by its status, α_i, where

$$\alpha_i = \begin{cases} 1 & \text{machine } i \text{ is up} \\ 0 & \text{machine } i \text{ is under repair} \end{cases} \tag{1}$$

This leads to the simple Markov chain shown in Figure 14. Here the transition from one state to another is related to the probability of failure (or repair). These probabilities will become the state transition probabilities.

Now consider the state of the transfer line. It consists of the status of each machine plus the amount of material present in the system. Specifically, the state of the transfer line is

$$s = (n, \alpha_1, \alpha_2), \tag{2}$$

where n = the number of pieces in the buffer plus the number of pieces in machine 2. The probability that the system is in state s, is

$$p(n, \alpha_1, \alpha_2). \tag{3}$$

Since $n, \alpha_1,$ and α_2 are all integers, the system is described by a set of mutually exclusive and collectively exhaustive states $s_1, s_2, \ldots s_m$. The system can only be in one of these states at any given instant of time.

The system may undergo a change of state (or a state transition) at discrete instants of time according to a set of probabilities. Let

Figure 13: A Simple Transfer Line

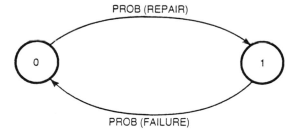

Figure 14: Markov Chain for a Single Machine

$P[s_i(k)]$ = probability that the system is in state s_i at time k.

Now we can say that a change of state will occur with probability,

$$P[s_j(k)|s_a(k-1),\ s_b(k-2),\ s_c(k-3)\ \ldots]$$
$$1 \leq j,\ a,\ b,\ c\ \leq\ m \qquad (4)$$
$$k\ =\ 1,\ 2,\ 3,\ \ldots$$

which is referred to as the transition probability.

B. THE MARKOV CONDITION

If $P[s_j(k)|s_a(k-1),\ s_b(k-2),\ s_c(k-3)\ \ldots]\ =\ P[s_j(k)|s_a(k-1)]$ for all $k, j, a, b, c, \ldots$ then the system is a discrete-state discrete transition Markov process. The implication of this condition is that the history of the system prior to its arrival at a has no affect on the transition to j. In a sense, the system has no memory.

C. STATE TRANSITION PROBABILITIES

For a Markov chain, we define the state transition probabilities as

$$p_{ij}\ =\ P[s_j(k)|s_i(k-1)],\ \ 1\ \leq\ i,\ j\ \leq\ m \qquad (5)$$

and each p_{ij} is independent of k. These probabilities can be included in a state transition matrix.

$$P\ =\ \begin{pmatrix} p_{11} & p_{12} & \cdots & p_{1m} \\ p_{21} & p_{22} & \cdots & p_{2m} \\ \vdots & \vdots & & \vdots \\ p_{m1} & p_{m2} & \cdots & p_{mm} \end{pmatrix} \qquad (6)$$

Also note that the state transition probabilities must satisfy

$$0\ \leq\ p_{ij}\ \leq\ 1 \qquad (7)$$

and

$$\sum_{j=1}^{m} p_{ij}\ =\ 1,\ \ i = 1, 2, \ldots m \qquad (8)$$

because the set of states is mutually exclusive and collectively exhaustive.

At this point, we would like to know what is the probability that the system will be in state s_i after k transitions, assuming that the state at $k = 0$ is known?

We can answer this question by finding a difference equation that relates the next state to the present state. We start with

$$P[s_j(k+1)] = P[s_1(k)]p_{1j} + P[s_2(k)]p_{2j} + \ldots P[s_m(k)]p_{mj} \qquad (9)$$

where $P[s_j(k+1)]$ is the probability that the system is in state s_j at time $k+1$. This equation represents all the possible ways that the system could have made the transition to state j, including staying there, i.e., including the term $P[s_j(k)]p_{jj}$. There are m of these equations since $j = 1, 2, \ldots m$. If we write out all m of them, we can put them in matrix form and obtain

$$(P[s_1(k+1)] \ P[s_2(k+1)] \ \ldots \ P[s_m(k+1)]) = \qquad (10)$$

$$(P[s_1(k)] \ P[s_2(k)] \ \ldots \ P[s_m(k)]) \begin{pmatrix} p_{11} & p_{12} & \cdots & p_{1m} \\ p_{21} & p_{22} & \cdots & p_{2m} \\ \vdots & \vdots & & \vdots \\ \vdots & \vdots & & \vdots \\ p_{m1} & p_{m2} & \cdots & p_{mm} \end{pmatrix} \qquad (11)$$

or

$$P(k+1) = P(k) \ P \qquad k = 0, 1, 2, \ldots \qquad (12)$$

To answer the original question, we solve this matrix difference equation by induction,

$$\begin{aligned} P(1) &= P(0)P \\ P(2) &= P(1)P = P(0)P^2 \\ &\vdots \\ &\vdots \end{aligned} \qquad (13)$$

which results in

$$P(k) = P(0) \ P^k \qquad k = 0, 1, 2, \ldots \qquad (14)$$

Example: A Serial Manufacturing System [20]

Figure 15 shows four stages of a serial manufacturing system. After each stage, the work-in-process is inspected. The scrap rates are shown in Table 3.

The problem is to obtain a Markov model for this system, find the state transition matrix, and a) find the probability that incoming material becomes finished product and b) find the probability that incoming material becomes scrap.

From the description of the system we can assign the following states:

s_1 incoming material
s_2 turning operation
s_3 milling operation
s_4 drilling operation
s_5 finished product
s_6 scrap

This leads to the model shown in Figure 16. From the scrap rates we can determine the transition probability matrix as

$$P = \begin{pmatrix} 0 & .998 & 0 & 0 & 0 & .002 \\ 0 & 0 & .97 & 0 & 0 & .03 \\ 0 & 0 & 0 & .95 & 0 & .05 \\ 0 & 0 & 0 & 0 & .98 & .02 \\ 0 & 0 & 0 & 0 & 1.0 & 0 \\ 0 & 0 & 0 & 0 & 0 & 1.0 \end{pmatrix} \qquad (15)$$

We can use this to find the probability that incoming material becomes finished product. This is the probability that the system is in state s_5 after 4 stages of operations. In other words, what is

$$P[s_5(4)] \text{ given that } P(0) = (1\ 0\ 0\ 0\ 0\ 0)? \qquad (16)$$

This is found from

$$P(4) = P(0)\,P^4 \qquad (17)$$

where $P[s_5(4)]$ is the fifth element in $P(4)$. An alternate solution is possible here since there are a limited number of paths. By inspection of the state diagram,

$$P[s_5(4)] = p_{12}\,p_{23}\,p_{34}\,p_{45} = (.998)(.97)(.95)(.98) = .90126386 \qquad (18)$$

Similarly, we can find the probability that incoming material becomes scrap, $(P[s_6(4)])$, from

$$P(4) = P(0)\,P^4 \qquad (19)$$

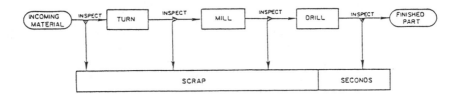

Figure 15: A Serial Manufacturing System [20]

Table 3: Operating and Scrap Rates

Incoming	Turn	Mill	Drill
	25 parts/hr	25 parts/hr	25 parts/hr
0.2%	3%	5%	2%

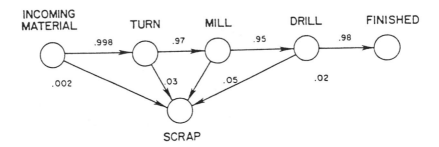

Figure 16: State Diagram for the Serial Manufacturing System

where $P[s_6(4)]$ is the sixth element in $P(4)$. But again, by inspection of the state diagram we can enumerate all the possible paths to s_6 to find,

$$
\begin{aligned}
P[s_6(4)] &= .002 + .998(.03) + .998(.97)(.05) \\
&\quad + .998(.97)(.95)(.02) \\
&= .09873614
\end{aligned}
\tag{20}
$$

D.   STATE CLASSIFICATION

Limiting State Probabilities

If as $k \to \infty$, $P(k)$ approaches a constant vector then the limiting state probabilities exist and are independent of the initial condition.

Let π be a 1 x m vector defined as

$$
\lim_{k \to \infty} P(k) = \pi
\tag{21}
$$

then π must satisfy

$$
\pi = \pi P
\tag{22}
$$

and the constraint $\sum_i^m \pi_i = 1$. In the previous example this yields $\pi_1 = \pi_2 = \pi_3 = \pi_4 = 0$ and $\pi_5 + \pi_6 = 1$. Therefore, no unique π can be found. Intuitively this makes sense since the amount of finished product depends on the initial state.

Transient State

State s_i is a transient state if you can leave the state but never return to it. In Figure 16, s_1, s_2, s_3, and s_4 are transient states.

Trapping State

State s_i is a transient state if the system enters the state and remains there. In Figure 16, s_5 and s_6 are trapping states.

VI. CONTINUOUS TIME MARKOV CHAINS

Up until now we have been concerned with discrete-state discrete-transition Markov chains where the transitions between states occur at uniformly spaced intervals of time. We would like to be able to handle transitions that occur at random time intervals because we would like to model random processing times, time between failures, and repair time. This requires the theory of continuous time Markov chains.

A. TRANSITION RATES AND THE STATE PROBABILITIES

In continuous time Markov chains, we deal with transition rates rather than state transition probabilities.

Definition: a_{ij} is the transition rate of a process from state i to state j, $i \neq j$. A is the transition rate matrix.

Definition: Given that the present state is s_i, the conditional probability that s_i goes to s_j during the time interval dt is $a_{ij}dt$. (Note that a_{ij} has units of sec^{-1} and so $1/a_{ij}$ = expected time between transitions.)

Let $P[s_i(t)]$ = probability that the system occupies state i at time t.   (23)

In this continuous time case, the problem is to find the probability of being in state j at time $t + dt$.

First we note that

$$P[s_j(t + dt)] = P[s_1(t)] a_j dt + \ldots P[s_{j-1}(t)] a_{j-1,j} dt +$$
$$P[s_j(t)] \text{ (probability of making no transition from } j\text{)} \qquad (24)$$
$$+ P[s_{j+1}(t)] a_{j+1,j} dt + \ldots P[s_m(t)] a_{mj} dt$$

Next, the probability of making no transition from j = 1− probability of making a transition from $j = 1 - \sum_{i \neq j}^{m} a_{ji} dt$.

Substituting and rewriting yields,

$$P[s_j(t + dt)] = P[s_j(t)](1 - \sum_{i \neq j}^{m} a_{ji} dt) + \sum_{i \neq j}^{m} P[s_i(t)] a_{ij} dt$$
$$j = 1, 2, \ldots m \qquad (25)$$

The goal is to find a solution to this equation. First, define the diagonal elements of the transition rate matrix A by,

$$a_{jj} = -\sum_{i \neq j}^{m} a_{ji} \qquad (26)$$

then

$$P[s_j(t + dt)] = P[s_j(t)] (1 + a_{jj}dt) + \sum_{i \neq j}^{m} P[s_i(t)] a_{ij}dt \qquad (27)$$

or

$$P[s_j(t + dt)] - P[s_j(t)] = \sum_{i=1}^{m} P[s_i(t)] a_{ij}dt \qquad (28)$$

Dividing by dt and taking the limit leads to

$$\frac{d}{dt} P[s_j(t)] = \sum_{i=1}^{m} P[s_i(t)] a_{ij} \quad j = 1, 2, \ldots m \qquad (29)$$

Next define

$$P(t) = (P[s_1(t)]\ P[s_2(t)]\ \ldots\ P[s_m(t)]) \qquad (30)$$

This results in the matrix differential equation

$$\frac{dP(t)}{dt} = P(t)\ A \qquad (31)$$

which has solution

$$P(t) = P(0)\ e^{At} \qquad (32)$$

If we are interested only in the steady state solution, then solve

$$0 = \pi\ A \qquad (33)$$

where π is a row vector of steady state probabilities.

Example: Transition Rates and the Exponential Distribution [15]

Consider a single machine M_1 that may be working (state 1) or not working (state 2). If it is working, there is a probability $5dt$ that it will break down in a short interval dt; if it is not working, there is a probability $4dt$ that it will be repaired in dt.

The problem is to find the probability that the machine will be operating at time t, if it is operating at $t = 0$.

Remark: $5dt$ and $4dt$ correspond to $a_{ij}dt$. They are equivalent to saying that the operating time between breakdowns is an exponentially distributed random variable with mean $1/5$ $(1/a_{ij})$, while the time for repair is exponentially distributed with mean $1/4$. This can be seen by examining the properties of the exponential distribution, given by

$$p(\tau) \;=\; \frac{1}{\bar{\tau}} \exp(-\tau/\bar{\tau}) \tag{34}$$

Here, τ is said to be exponentially distributed with mean $= E(\tau) = \bar{\tau}$. The following points can be summarized for the exponential distribution:

- The random variable in the exponential distribution is the time interval between adjacent events.

- The average time interval between adjacent events is $\bar{\tau}$.

- The rate at which events occur is $1/\bar{\tau}$.

Solution: Let the time between failures (time under repair) be an exponentially distributed random variable with rate $p_i(r_i)$. M_1 can then be modeled as in Figure 17.

First, by inspection of the model we can write,

$$A \;=\; \begin{pmatrix} a_{11} & 5 \\ 4 & a_{22} \end{pmatrix} \tag{35}$$

The diagonal elements in the A matrix are given by,

$$a_{11} \;=\; -\sum_{i\neq 1}^{2} a_{i2} \;=\; -a_{12} \;=\; -5 \tag{36}$$

$$a_{22} \;=\; -\sum_{i\neq 2}^{2} a_{2i} \;=\; -a_{21} \;=\; -4 \tag{37}$$

so that

$$A \;=\; \begin{pmatrix} -5 & 5 \\ 5 & -4 \end{pmatrix} \tag{38}$$

Next, solve $P(t) = P(0)e^{At}$ where $P(0)$ has been given as $(1\ 0)$. This yields,

$$P(t) \;=\; (4/9 \;\; 5/9) + e^{-9t}\,(5/9 \;\; -5/9) \tag{39}$$

In steady state we have

$$\pi \;=\; (4/9 \;\; 5/9) \tag{40}$$

which says that the machine spends most of its time in the repair state.

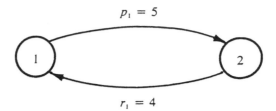

$$p_1 = 5$$

$$r_1 = 4$$

Figure 17: Single Machine Model

VII. STOCHASTIC PETRI NETS

Stochastic Petri nets (SPNs) are timed Petri nets in which the transition times are random variables. If the transition times have exponentially distributed firing rates, then the SPN is equivalent to a continuous time Markov chain [7].

Generalized stochastic Petri nets (GSPNs) incorporate both timed transitions (drawn as white boxes) and immediate transitions (drawn as thin black bars). The timed transitions have an exponentially distributed firing rate λ and fire $1/\lambda$ time units after being enabled. The immediate transitions fire in zero time.

Example: GSPN Model for an Unreliable Machine [19,22]

Consider a machine characterized by a processing rate μ, a failure rate p, and a repair rate r. Each represents the firing rate of a timed transition described by an exponential distribution.

First, we identify the following places:

p_{uf} - machine is up and free

p_{ub} - machine is up and busy

p_d - machine is down

p_{ui} - machine is up and idle

The following events serve to identify the transitions:

t_i - material is input to the system

t_f - machine has failed

t_r - machine is being repaired

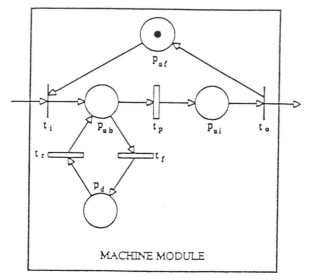

Figure 18: GSPN Model of an Unreliable Machine

t_p - a part is being processed

t_o - finished parts are dispatched from the machine

The GSPN model is shown in Figure 18. Note that t_i and t_o are immediate transitions. The initial marking indicates that the machine is up and free. Assuming that the machine is never starved, t_i fires immediately, and a token is placed in p_{ub}. Next, t_f or t_p may fire depending on the firing rates. If t_f fires, the machine goes down, indicated by the token in p_d. After an average time of $1/t_r$. the machine is repaired and is up and busy again. The machine may fail again, but eventually a part is finished and the machine becomes up and idle. Assuming that the machine is never blocked, t_o fires in zero time and the machine is up and free again.

A.   PETRI NETS AND MARKOV CHAINS

We can introduce the concept of state in a Petri net by considering each possible token distribution as a state. This can be illustrated by considering the previous example where the state can be defined as the number of tokens in $(p_{uf}, p_{ub}, p_{ui}, p_d)$.

We start with the initial marking $m_o = (1\ 0\ 0\ 0)$. This enables t_i and generates the new marking (or state), (0 1 0 0). From this state we can go to (0 0 0 1) by firing t_f or to (0 0 1 0) by firing t_p. In (0 0 0 1) we would return to (0 1 0 0) by firing t_r. We would leave (0 1 0 0) by firing t_p to enter the new state (0 0 1 0). After firing t_o we would return to m_o. Figure 19 shows the state diagram.

Because the system contains immediate transitions, we actually spend zero time in those states. This requires some additional concepts.

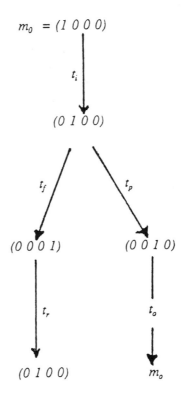

Figure 19: State Diagram of an Unreliable Machine

B.   VANISHING AND TANGIBLE MARKINGS

Markings (or states in Figure 19) in which at least an immediate transition is enabled are called vanishing markings. On the other hand, markings in which only exponential transitions are enabled are called tangible markings.

A chain (like the one in Figure 19) that contains tangible and vanishing (markings) states is called an embedded Markov chain. If we eliminate the vanishing states from Figure 19, we have a chain that contains only tangible states and it is called the reduced embedded Markov chain.

Finally, we state some important results for GSPNs. The solution for the steady state probability distribution may or may not exist. This depends on the properties of the underlying Petri net. The underlying Petri net is the net that results by treating immediate and exponential transitions as if they were the same. Then this net is analyzed and the steady state probability distribution exists if

- the underlying net is bounded,

- the initial marking is reachable from all reachable markings, and

- the firing rates are exponentially distributed random variables.

Example: Markov Chains for the Unreliable Machine

Figure 19 contains both vanishing and tangible states. Therefore, it is an example of an embedded Markov chain.

States (0 0 1 0) and (1 0 0 0) are vanishing states; zero time is spent in these states. Removing these states results in Figure 20 and it represents the reduced embedded Markov chain. From here, we proceed to find the steady state probability of being in state π_1 and π_2. We follow exactly the same procedure that we described for continuous time Markov chains.

From the reduced embedded Markov chain we find the transition rate matrix as

$$A = \begin{pmatrix} -t_f & t_f \\ t_r & -t_r \end{pmatrix} \qquad (41)$$

Now suppose $t_f = 3$, $t_r = 5$, $t_p = 10$, and we wish to find 1) probability that the machine is down, and 2) the production rate of the machine.

1. The probability that the machine is down is π_2. So we need to solve

$$0 = (\pi_1 \ \pi_2) \begin{pmatrix} -3 & 3 \\ 5 & -5 \end{pmatrix} \qquad (42)$$

subject to $\pi_1 + \pi_2 = 1$. This yields $\pi_1 = 5/8$ and $\pi_2 = 3/8$.

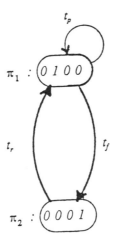

Figure 20: Reduced Embedded Markov Chain of
an Unreliable Machine

2. To find the production rate of the machine, we have to use the steady
   state probabilities in conjunction with other information we have about
   the system. Specifically, the production rate, P = (probability that the
   machine is up) (rate at which parts can be produced) = $\pi_1\, t_p$ = 6.25
   parts/unit time.

C. AUTOMATING THE MARKOV ANALYSIS PROCESS

The previous examples have illustrated the basic steps that are used to go from
the Petri net model to the reduced embedded Markov chain. These steps can
be summarized as follows:

1. Develop a Petri net model for the system that takes into account the
   relative duration of the operations. In other words, immediate transitions
   can be used to model relatively short operations.

2. Check the net for liveness and boundedness.

3. From the initial marking, generate the state space that includes both
   vanishing and tangible markings.

4. Remove the vanishing markings to obtain the state space for the reduced
   embedded Markov chain. Assign states $(\pi_1, \pi_2, \ldots, \pi_m)$ = π.

5. Associate transition firing rates with state transition rates to build the
   transition rate matrix, A.

6. Solve the set of linear algebraic equations 0 = $\pi\, A$ for the steady state
   probabilities.

7. Calculate the desired performance measures from these probabilities. Note that most performance measures are a function of these steady state probabilities, e.g., production rate in the previous example.

These steps have been automated in several software packages.

Chiola [23] has developed GreatSPN for the construction and analysis of complex generalized SPN models. This software accepts deterministic delays or exponentially distributed firing rates. It also computes the transient and steady-state solutions to the Markov chains.

Dugan et al. [24] have developed the Duke extended SPN evaluation package (DEEP) for the performance analysis of SPN models. This has led to a more recent version [25]. Holiday and Vernon [26] have developed the GSPN analyzer for the performance evaluation of generalized timed Petri net models.

It should be emphasized, that these software packages are *not* simulations of Petri net models. They are solving the equivalent Markov chain. The value of Petri nets lies in the fact that they are conceptually easy to use as a modeling tool, yet they can represent very large (and equivalent) Markov chains.

Example: A Three Machine Two Buffer Transfer Line [22]

Figure 21 shows the GSPN model for a three machine two buffer transfer line. This model includes the use of an inhibitor arc as part of each buffer model. At buffer 1, when k_1 tokens accumulate in p_5, the inhibitor arc prevents transition t_5 from firing, even if there is a token in p_4. t_5 will not fire again until there is less than k_1 tokens in p_5. In other words, there must be a buffer vacancy before t_5 can put a part in the buffer. The inhibitor arc is a convenient modeling tool but it can be replaced by simple places and transitions (Figure 22). Thus, the equivalence with Markov chains is not destroyed when using inhibitor arcs.

This is a good example of how Petri nets manage the complexity of Markov chains. In this example, the equivalent Markov chain has 318 states and 1281 arcs.

The following performance measures can be computed for this model:

$$
\begin{aligned}
\text{Average production rate } P &= \text{rate}(t_2)\text{prob } \{t_2 \text{ is enabled}\} \\
&= \text{rate}(t_7) \text{ prob}\{t_7 \text{ is enabled}\} \\
&= \text{rate}(t_{12}) \text{ prob}\{t_{12} \text{ is enabled}\}
\end{aligned}
$$

Average in-process inventory $n = E[M(p_2) + M(p_3) + M(p_4) + M(p_5) + M(p_7) + M(p_8) + M(p_9) + M(p_{10}) + M(p_{12}) + M(p_{13}) + M(p_{14})]$

Average machine utilization

$$UM_1 = \text{prob}\{M(p_2) = 1\}$$

$$UM_2 = \text{prob}\{M(p_7) = 1\}$$

$$UM_3 = \text{prob}\{M(p_{12}) = 1\}$$

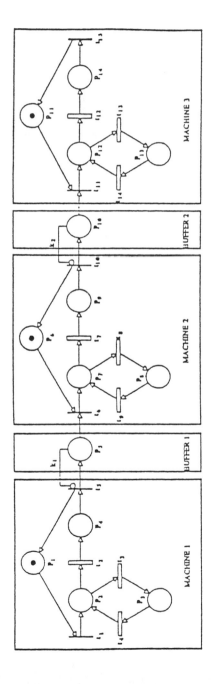

Figure 21: GSPN Model of a Three Machine Two Buffer Transfer Line

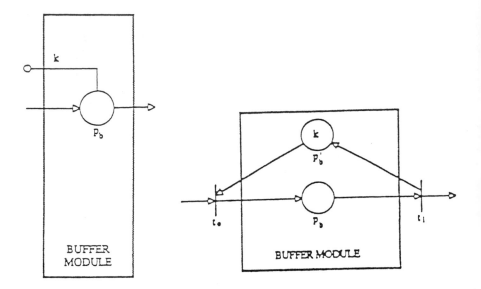

Figure 22: Equivalent Representation of An Inhibitor Arc

Average buffer utilization

$$UB_1 = \text{prob}\{M(p_5) \neq 0\}$$

Probability of machine blocked

$$B_2 = \text{prob}\{M(p_9) = 1\}$$

Probability of machine starved

$$S_2 = \text{prob}\{M(p_{11}) = 1\}$$

The results were obtained for buffer sizes $k_1 = 6$ and $k_2 = 3$. For machine 1, $\mu_1 = p_1 = r_1 = 1$ and for machines 2 and 3 the processing rate $= 10$, failure rate $= 0.1$ and the repair rate $= 10$.

Figure 23 shows the effect on production rate when μ_1 is varied between 0.1 and 1000 ($P(1)$). $P(2)$ is for μ_2 varying between 0.1 and 1000 and $P(3)$ is the similar case for μ_3. A similar interpretation applies to Figure 24. Each point in these figures represents the solution to a specific Markov chain as a parameter is varied. The graphs show what happens under different processing conditions and can answer a lot of "what if"questions about the system. For example, Figure 23 shows that increasing μ_1 beyond approximately 30 will not increase $P(1)$. Thus, investing in a faster machine cannot be justified. Additional results can be found in [22,27].

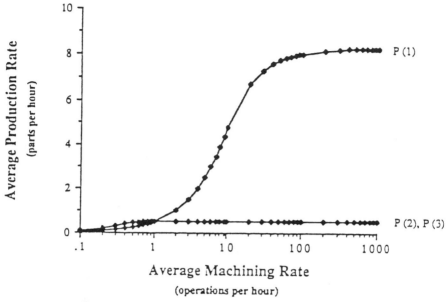

Figure 23: Production Rate vs. Machining Rate

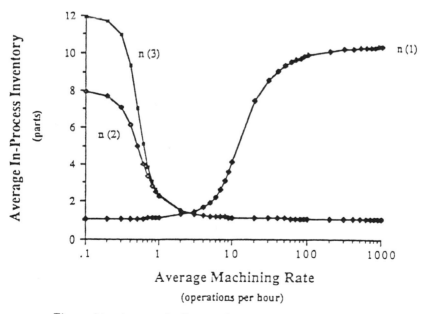

Figure 24: Average In-Process Inventory vs. Machining Rate

VIII. CONCLUSION

This chapter has presented and illustrated some of the advantages of using Petri nets for the modeling, analysis and control of manufacturing systems. Figure 25 organizes the applications presented and presents Petri nets as a single representation tool. Nets can aid in modeling, validation and performance evaluation at design time. Once the system shows desirable behavior, the net can be translated to control and monitor operations at run time.

The following list cites the advantages the authors have found in using Petri nets:

1. Ease of systems modeling

     Excellent visualization of system dependencies

     Focus on local information

     Top down (stepwise refinement) design

     Bottom up (modular composition) design

2. Ability to generate control code directly from the graphical representation.

3. Ability to validate code by mathematically based computer analysis - no time consuming simulations.

4. Performance analysis without simulation possible.

5. Simulation can be driven from the model.

6. Status information from token player allows for real time monitoring.

It is hoped that through this rather brief tutorial many more control system engineers will consider this tool for the synchronization and coordination of their plants.

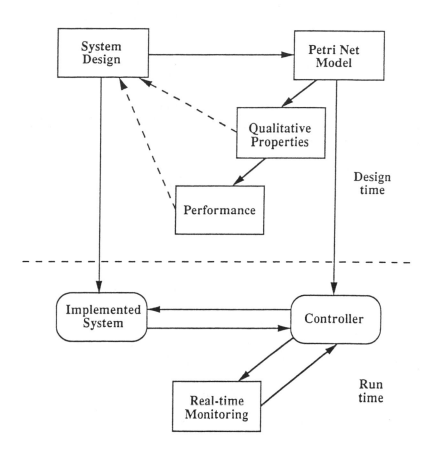

Figure 25. Petri Nets as a Single Representation for
Modeling, Analysis, and Control

IX.   References

1. T. Murata, "Petri nets: properties, analysis and applications," *Proc. IEEE*, Vol. 77, No. , Apr. 1989.

2. J.L. Peterson, *Petri Net Theory and The Modeling of Systems*, Prentice Hall, 1981.

3. K. Lautenbach, "Linear algebraic techniques for place/transition nets," *Advances in Petri nets 1986 (Lecture Notes in Computer Science 254)*, Springer-Verlag, pp. 142-167, 1986.

4. Y. Narahari and N. Viswanadham, "A Petri net approach to the modeling and analysis of flexible manufacturing systems," *Annals of Operations Research*, Vol. 3, pp. 449-472, 1985.

5. Silva, M. *Las redes de Petri en la Automatica y la Informatica*, Editorial AC, Madrid, 1985.

6. G. Berthelog, "Transformations and decompositions of nets," *Advances in Petri nets 1986 (Lecture Notes in Computer Science 254)*, Springer-Verlag, pp. 359-376, 1986.

7. M.D. Jeng and F. DiCesare, "A review of synthesis techniques for Petri nets," *Proc. 2nd International Conf. CIM*, Rensselaer Polytechnic Institute,Troy, NY, May 1990.

8. Zhou, M.C. and F. DiCesare, "Parallel and sequential mutual exclusions for Petri net modeling for manufacturing systems," *IEEE Trans. on Robotics and Automation*, Vol. 7, No. 7, pp. 515-527.

9. D.L. Rudolph, *Petri Net-based control of a Flexible Manufacturing System*, Master's Thesis, Rensselaer Polytechnic Institute, Troy, New York, 1989.

10. R. Valette, M. Courvoisier, H. Demmou, J.M. Bigou, C. Desclaux, "Putting Petri nets to work for controlling flexible manufacturing systems", *Proc. of Int. Symp. on Circ. & Sys.*, pp. 929-932, Kyoto, Japan, 1985.

11. T. Murata, N. Komoda,and K. Matsumoto, "A Petri net based controller for flexible and maintainable sequence control and its applications in factor automation," *IEEE Trans. on Industrial Electronics*, IE-33, 1-8, 1986.

12. M. Silva, and S. Velilla, "Programmable logic controller and Petri nets: a comparative study," *Proc. of the IFAC Conf. on Software for Computer Control*, pp. 38-88, Madrid, Spain, 1982.

13. D. Crockett, A. Desrochers, F. DiCesare, and T. Ward, "Implementation of a Petri net controller for a machining workstation," *Proc. of IEEE Robotics and Automation Conference*, Raleigh, North Carolina, pp. 1861-1867, 1987.

14. M.C. Zhou, F. DiCesare, and D. Rudolph, "Control of a flexible manufacturing system using Petri nets," in *1990 IFAC Congress Conference*, July 1990.

15. R.A. Howard, *Dynamic Programming and Markov Processes*, M.I.T. Press, Cambridge, MA, 1960.

16. R.A. Howard, *Dynamic Probabilistic Systems*, Vol. 1 (Markov Models) and Vol. II (Semi-Markov and Decision Processes), John Wiley and Sons, Inc., New York, 1971.

17. L. Kleinrock, *Queueing Systems, Volume 1: Theory*, John Wiley and Sons, Inc., New York, 1975.

18. S.M. Ross, *Introduction to Probability Models*, 2nd edition, Academic Press, Orlando, Florida, 1980.

19. A.A. Desrochers, *Modeling and Control of Automated Manufacturing Systems*, IEEE Computer Society Press, Washington, DC, 1990.

20. R.P. Davis and W. Kennedy, "Markovian Modeling of Manufacturing Systems", *International Journal of Production Research*, 1987, Vol. 25, No. 3, pp. 337-351.

21. M.K. Molloy, "Performance Analysis Using Stochastic Petri Nets", *IEEE Transactions on Computers*, Vol. 1-31, No. 9, September 1982, pp. 913-917.

22. R.Y. Al-Jaar and A.A. Desrochers, "Performance Analysis of Automated Manufacturing Systems Using Generalized Stochastic Petri Nets", *IEEE Transactions on Robotics and Automation*, December 1990, pp. 621-639.

23. G. Chiola, "A Graphical Petri Net Tool for Performance Analysis", *Proceedings of the 3rd International Workshop on Modeling Techniques and Performance Evaluation*, AFCET, Paris, France, March 1987.

24. J.B. Dugan, A. Bobbio, G., Ciardo, and K. Trivedi, "The Design of a Unified Package for the Solution of Stochastic Petri Net Models", *Proceedings of the IEEE International Workshop on Timed Petri Nets*, Torino, Italy, July 1985, pp. 6-13.

25. G. Ciardo, "Manual for the SPNP Package", Duke University, Durham, NC, July 1988.

26. M.A. Holliday and M.K. Vernon, "A Generalized Timed Petri Net Model for Performance Analysis", *Proceedings of the IEEE International Workshop on Timed Petri Nets*, Torino, Italy, July 1985, pp. 181-190.

27. R.Y. Al-Jaar and A.A. Desrochers, "Modeling and Analysis of Transfer Lines and Production Networks Using Generalized Stochastic Petri Nets", in *Proc. Conf. Univ. Programs Comput. Aided Eng. Design Manuf.*, (Atlanta), June 27-29, 1988, pp. 12-21.

EVALCON: A SYSTEM FOR EVALUATION OF CONCURRENCY IN ENGINEERING DESIGN

ANDREW KUSIAK
EDWARD SZCZERBICKI

Intelligent Systems Laboratory
Department of Industrial Engineering
The University of Iowa
Iowa City, IA 52242

I. INTRODUCTION

Design undergoes evaluation at all its stages. Evaluation of design from the perspective of various life-cycle attributes, such as manufacturability, assemblability, reliability, and so on, is named a concurrency evaluation. The concurrency evaluation process in routine design should assist a designer in specification, selection, and synthesis of parts. The main difficulty associated with the evaluation process is caused by the incompleteness and uncertainty of information. To facilitate the routine design and to allow for incremental concurrency evaluation as the design evolves, an interactive design support system is proposed. The approach presented employs concepts from cluster analysis, decision science, information theory, and artificial intelligence. An important part of this approach is an existing design library which includes parts, subassemblies, and the corresponding functional specifications. Parts and functions are grouped using cluster analysis. The evaluation of concurrency attributes at the specification stage is supported by a minimax optimization algorithm. In the part selection phase, the Dempster-Shafer approach is used. The Bayesian inference assists the designer in concurrency evaluation in the synthesis phase. Transformation from conceptual to embodiment design is supported by graph-theoretic tools. The main novel characteristics of the presented system are as follows:

(a)   it offers a designer three different approaches for the evaluation process that are suitable for different levels of uncertainty and availability of information,

(b)   it is embedded in an existing design library and reduces the complexity of part selection by using cluster analysis,

(c)   it assists a designer in transformation from conceptual to embodiment design.

All ideas introduced in the chapter are illustrated with examples. An object programming implementation is also briefly discussed.

A. PRINCIPLES OF A FORMAL APPROACH TO DESIGN

Design process, in its broad meaning, has to take into account not only economic, manufacturing, and marketing considerations but also social, human, and even political ones [1]. Bearing in mind this important aspect, one may recognize that there are two basic aims of the design process:

(i)   to create a product that satisfies material needs,

(ii)   to create a knowledge about man-environment systems that satisfies
     intellectual needs.

The creation context in the two above cases is driven by the difference between "what is" and "what is desired". Design attempts to affect this difference (ΔS) that can be expressed as:

$$\Delta S = |S_0 - S| \tag{1}$$

where:

S   the actual state and design capabilities of the surrounding world,

S_0 material and intellectual needs.

The relation between recognized needs, design process, and design modeling is presented in Figure 1.

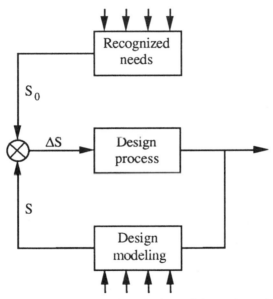

Figure 1. Design process and its relation with needs and modeling

The relation described by Eq. (1) and depicted in Figure 1 is dynamic. The design process must adjust to the changing needs S_0 and changing capabilities S. Thus, introducing t as time one has the following:

$$\begin{aligned} S_0 &= S_0(t) \\ S &= S(t) \\ \Delta S &= \Delta S(t) \end{aligned} \tag{2}$$

For the the general case described by Eq. (2), two specific cases of affecting the difference in Eq. (1) can be discussed. The first one is described as:

$$S_0(t) = const$$

$$S = S(t)$$

(3)

which means that for a given time period $\Delta t = t_2 - t_1$ material and intellectual needs do not change. Design capabilities, however, do change and the design process is thus expressed by:

$$\Delta S(t) = |const - S(t)| \text{ ---> min} \qquad (4)$$

In the second case one has:

$$S(t) = const$$

$$S_0 = S_0(t)$$

(5)

which means that for a given period of time, design capabilities are the same but the needs change and thus the design process is given as:

$$\Delta S(t) = |S_0(t) - const| \text{ ---> min} \qquad (6)$$

The above specific cases may happen for relatively short time periods (they may be caused by, for example, economic stagnation). In the long-run, however, situation described by Eq. (2) usually takes place. The smaller the difference ΔS between the needs and capabilities the better it is from the society's point of view. This difference is constrained by the cost and time. The hypothetical relationship between time and cost constraints and ΔS is nonlinear as presented in Figure 2.

For the dependency illustrated in Figure 2, there exists a point from which the gradient of cost and time increase is larger than the relating gradient of ΔS

decrease. This indicates why designers should concentrate their efforts not on satisfying the relation ΔS ---> 0, but rather on ΔS ---> min.

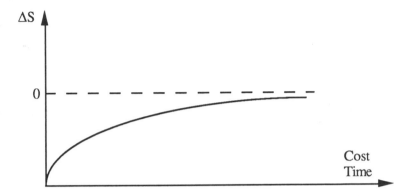

Figure 2. Hypothetical relationship between ΔS and constraints created by cost and time

The research in engineering design concentrates mainly on design modeling (see Figure 1). Various models of the design process have been developed. A model that is widely agreed upon was proposed by Asimow [2]. Coyne *et al.* [3] discussed the Asimow's model from the knowledge-based perspective. The model includes three phases: analysis, synthesis, and evaluation. Since the model was developed, the interpretation of its three phases has undergone changes. Nowadays, it is realized that the design phases should not be addressed as a pure hierarchy or ideal sequential order. Rather, they often mix and are carried out simultaneously. This is especially true when the evaluation phase is seen as the one that evaluates the satisfaction of concurrency attributes.

B. MODEL OF INFORMATION FLOW IN DESIGN SYSTEMS

In the last two decades, attempts have been made to formalize the design process. Various methodologies have been developed. For example, a decision-making approach was proposed in [4] and recently expanded in [5]. System analysis and hierarchical decomposition was applied to design by

Mesarovic *et al.* [6]. An optimization approach was developed by Radford and Gero [7] for architectural design, and Arora [8] for mechanical design. Design process was presented as an information processing activity by Koomen [9] and Conant [10]. A knowledge-based approach was presented in [3]. Pahl and Beitz [11] and Hubka [12] proposed a systematic approach to engineering design.

There are two ways in which a theory can be developed. The first approach that can be called *a posteriori,* describes in a model phenomena observed over a period of time in existing real-life systems. The second approach (*a priori*) builds a model in an attempt to predict phenomena that have not taken place yet. Since many of the existing and working design systems are usually ahead of modeling that describes them, the design theory is built in an *a posteriori* way and its main purpose is to gain better understanding of the design process. The above nature of design theory makes it necessary to adjust it and incorporate new elements as real systems evolve. The complexity and creativity associated with engineering design has resulted in a number of implementations of artificial intelligence (AI) approaches in engineering design and related areas, for example:

(a)   design of printed circuit boards (Simoudis [13], Steinberg and Mitchell [14]),

(b)   design of computer software (Ellsworth *et al.* [15]) ,

(c)   architectural design (Flemming *et al.* [16], Gero and Coyne [17], Maher [18]),

(d)   structural design (Grierson and Cameron [19]),

(e)   engine design (Shen *et al.* [20]).

Input and output of existing models consists of signals that are being evaluated by a designer. As design systems evolve into CAD (computer aided design) and ICAD (intelligent computer aided design) in a new modeling approach an assumption should be made that some evaluation is done by the system itself that may be supported by artificial intelligence techniques. As a consequence, the model of a design system should consist of three groups of modules: modules controlled by a designer, controlled by techniques incorporated, and those of integrated control performed by both a designer and the system itself.

There is a tendency towards shortening a product life-cycle. This can be achieved mainly by the acceleration of information flow and thus models of design systems should include algorithms (an algorithm usually accelerates the flow of information compared to an ad hoc approach) and mechanisms that ease the adaptation to change in designed objects (high degree of flexibility should accompany the dynamics of the changing needs). This phenomena results in a modeling approach in which the design process is seen as an information processing activity. A number of authors have stressed the importance of information flow and uncertainty in design (see, for example [1, 9, 10, 21, 22]).

The complexity of the design process calls for modeling techniques based on a system approach. A system was defined for designers and engineers in [5] as "*a grouping of associated entities which is characterized by a mental construct; one of the associated entities is the boundary of the system*". This definition stresses, through the concept of the boundary, the importance of the environment of a system and (through the concept of association) the need for design process control, organization, and planning. Environment interacts with the design system in two ways. First, by intensity of the input of design resources (mainly information and energy) that can be measured in monetary scale (denoted by K). Second, by qualitative evaluation of the design effect (denoted by F) described by life-cycle characteristics of designed object (safety, serviceability, quality, etc.). Control, organization, and planning phenomena can be expressed by the efficiency of resources allocation (K/F) and by the efficiency of the system (F/K). Altogether, the generalized formal model of the design system (DS) can be expressed with the following four parameters:

$$DS = <K, F, e_1, e_2> \qquad\qquad (7)$$

where $e_1 = K/F$ and $e_2 = 1/e_1$.

This model suggests an approach in which the sensitivity of design system to changes of various design attributes can be studied. For example, suppose that K and F is expressed as a function of design attribute a_1 (in Section II attributes a are introduced as concurrency attributes):

$$e_2K(a_1) = F(a_1) \qquad (8)$$

From Eq. (8) one has:

$$[1/K(a_1)][dF(a_1)/da_1] - [F(a_1)/K(a_1)^2][dK(a_1)/da_1] = 0 \qquad (9)$$

and

$$[K(a_1)\Delta F(a_1) - F(a_1)\Delta K(a_1)]/K(a_1)^2 = 0 \qquad (10)$$

which can be expressed as

$$[K(a_1)/F(a_1)]\Delta F(a_1) = \Delta K(a_1) \qquad (11)$$

Equation (11) expresses sensitivity of the response of the design system to changes in attribute a_1. Any change in a_1 may affect Eq. (11) in two ways:

$$[K(a_1)/F(a_1)]\Delta F(a_1) > \Delta K(a_1) \qquad (12)$$
$$[K(a_1)/F(a_1)]\Delta F(a_1) < \Delta K(a_1) \qquad (13)$$

the first one (Eq. (12)) describing the desired direction of changes in design and the second one (Eq. (13)) the changes that are not desired.

The decision aspect is incorporated into the generalized model of the design process in the following way. Design can be seen as a series of decisions made by a designer. Each decision involves collecting and processing of information that is the basis for evaluation of design variants. Three general (global) criteria $c1$, $c2$, and $c3$ for evaluation are structured dynamically as functions of main groups of information (I_1 - information channel of design and engineering knowledge, I_2 - information channel of economic constraints, I_3 - information channel of recognized needs):

$$c_1 = c_1(I_1, t)$$
$$c_2 = c_2(I_2, t) \qquad\qquad (14)$$
$$c_3 = c_3(I_3, t)$$

Information flow in design seen as decision-making process is presented in Figure 3.

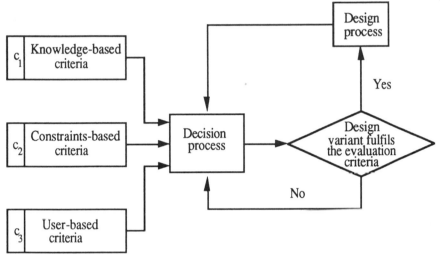

Figure 3. The model of information flow for design decision-making

The approach presented in this chapter creates the framework for design formalization and modeling. Such a framework is illustrated in Figure 4.

As shown in Figure 4, the overall information flow $I(t)$, where t denotes time, is first selected and then decomposed into channels $I_1(t)$, $I_2(t)$, ..., $I_n(t)$. Since the information may include some noise, the filtering block is included. The output of the design process is an engineering abstraction of the designed object which evaluation function is denoted by $F(t)$. The actual output $F(t)$ is compared with the required standard denoted by $F_0(t)$ in Figure 4. The difference between the two is minimized in a feedback loop with corrections as presented in Figure 4.

Design projects fall within one of three categories: original, adaptive, and variant [5]. The last category, that is named in this chapter a *routine* design,

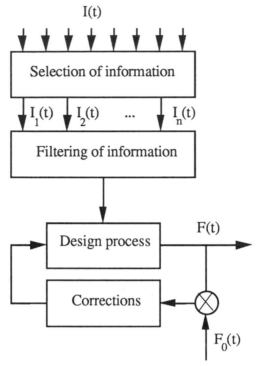

I(t)

Figure 4. Information flow in design

uses the parts existing in the design library. Since routine design prevails in practice, there is a need to formalize its activities and assist the designer during the process of choosing among the existing parts and synthesizing them into a designed object. The routine design might benefit in two ways. First, application of algorithms might shorten the design cycle. Second, the decision support techniques embedded in the system will help the designer to make better decisions and consequently to produce designs of better quality.

II. EVALUATION OF THE DEGREE OF CONCURRENCY SATISFACTION IN DESIGN

A product life cycle includes acquisition (conceptual and preliminary design, detailed design and development, and production), and utilization (product use, phaseout, and disposal) [23]. The life cycle incorporates into the design process various life-cycle objectives that in this chapter are called *concurrency attributes*. Concurrency attributes include esthetics, manufacturability, assemblability, reliability, serviceability, and so on. Concurrent design refers to the integration of various concurrency attributes within the broad scope of acquisition and utilization [24]. This means that a designed object undergoes an evaluation from different perspectives that represent various concurrency attributes. Such an evaluation is included in the model of the design process presented in Figure 5.

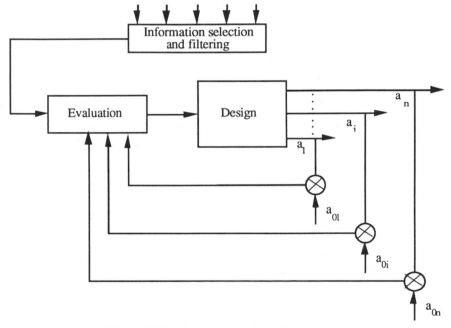

Figure 5. Design as a goal-seeking system

In Figure 5, a_1, a_2, ..., a_n denote the concurrency attributes used for evaluation of designs. Each attribute contributes to the overall evaluation function given as:

$$F = f(a_1, a_2, ..., a_n). \qquad (15)$$

For modeling purposes, evaluation of each attribute a_i is accomplished by comparing its value with the value of a desired state represented in Figure 5 by a_{0i} i=1, ..., n. The system in Figure 5 belongs to the class of goal-seeking systems for which the theoretic framework was developed by Mesarovic and Takahara [25]. A practical implementation of the goal-seeking system requires that measures for evaluation of an actual and desired state of a_i be established. As reported in the literature on design of systems for evaluation of concurrency attributes such a measure has to take into account uncertainty and incompleteness of the information available, and is partly based on a subjective judgement of a designer or an expert. For example, Ishii *et al.* [26] developed an evaluation measure of concurrency that uses the concept of design compatibility analysis. This measure is expressed as the simultaneous engineering index (SEI) valued between 0 and 1. The SEI index is compared for each concurrency attribute (serviceability, functionality, machinability, and assembly) with the threshold value defined by the user. Maher [27] proposed an evaluation method based on preference measure that is specified as weights for each criterion being considered. Wood *et al.* [28] argued that in engineering design there are subjective uncertainties that can be measured by a possibility function. Based on this measure an evaluation system for design with imprecise and incomplete design information was developed. Kota *et al.* [29] introduced the goodness-of-match for the evaluation of satisfaction of design constraints. A value of 1 denotes a perfect match and a value of 0 indicates no match at all. Zhang and Siddall [30] proposed a desirability function that evaluates design with respect to different performance criteria such as strength, durability, etc.

In this chapter, a notion of the *degree of satisfaction of concurrency attribute* is introduced. It can be expressed numerically, statistically, or on an ordinal scale. The degree of satisfaction of a given concurrency attribute simulates the judgement on how well a design (or its subset) satisfies the requirements associated with the corresponding attribute. Each design imposes different values on the degree of satisfaction of concurrency attributes. For example, a throw-away product may be assigned the required degree of serviceability satisfaction as 0 in the scale 0 to 1 (it does not need to be serviced), and the required degree of manufacturability as close to 1 (it must be inexpensive and therefore easily manufacturable). Design of a coupling has a low value of the degree of satisfaction of the aesthetic attribute, but a high value for reliability. In each case, the judgement of the degree of satisfaction of the concurrency attribute is at least partially subjective and is based on designer's knowledge. Depending on the availability and completeness of information, a degree of satisfaction is expressed in Section III of this chapter with the *regret* function, or minimum and maximum *likelihood* function, or *posterior probability*.

The uncertainty associated with information, effective uncertainty handling, and modeling of the decision making process based on available information become increasingly important in evaluation of the degree of satisfaction of concurrency attributes [5]. This evaluation, especially during the early stages of design, is based on a subjective assessment of relevant criteria. In the absence of complete information, methods that consider subjective criteria while dealing with uncertainty, imprecise, and incomplete information are suitable as design evaluation techniques. The latter has been realized by some researchers. Wilde [31] developed logical conditions for reliability, safety, cost, marketability, aesthetic, and environmental impact of the designed object. This concept was one of the first attempts to deal with design decision making under the concurrency constraints. The Wilde's approach, however, requires precise information as it uses symbolic logic. The relevance of imprecision and uncertainty in engineering design has been discussed in [28]. Their approach is embedded in the fuzzy set theory and is primarily focused on imprecision (uncertainty in choosing among alternatives) along with stochastic and possibilistic uncertainties. Maher [27] used the concept of Pareto optimality combined with preferences specified as weights for each criterion for identifying the best design solutions. A model for evaluation of designs in concurrent engineering using design compability analysis was developed in [26]. The above research sets a new trend in supporting of some design activities. It appears that this trend will continue as it shows some promise in dealing with those activities performed by a designer that are vague and uncertain in nature. The evaluation of the degree of satisfaction of concurrency attributes is without doubt one of such activities.

The research presented in this chapter differs from the existing research in one important way. It aims at developing a system that incorporates three different approaches that match the availability of concurrency relevant information, i.e., data on which a degree of satisfaction of concurrency attributes can be judged. At the early stages of design such information is usually not available and a system is needed to provide a rough evaluation of the degree of satisfaction of concurrency attributes. At later stages of design, as the amount of available information increases, techniques suitable for information-rich environment are more desirable. An architecture of the system that satisfies the above is presented in Figure 6.

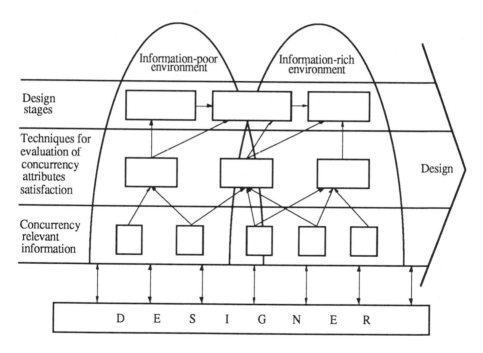

Figure 6. Architecture of the system for evaluation of the degree of satisfaction of concurrency attributes

A. CONCURRENCY ATTRIBUTES

For each concurrency attribute a set of *design features* can be considered. Depending on the domain and the type of the designed object, the designer may consider different concurrency attributes and the corresponding design features. An example list of concurrency attributes and design features is presented in Figure 7.

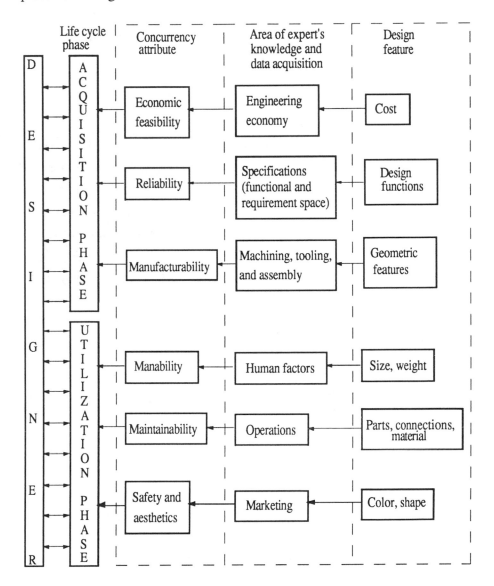

Figure 7. Sample concurrency attributes and the corresponding areas of expert's knowledge and design features

As shown in Figure 7, a designer has to satisfy a set of concurrency attributes. Each attribute is to be satisfied to a certain degree and it becomes apparent that the design system should include tools that will allow the designer to observe and control its value when a design decision is made. These tools should handle uncertain and incomplete information and be suitable for dealing with a subjective judgement.

III. SYSTEM FOR CONCURRENT DESIGN

The system for concurrent design assists the designer to consider various life-cycle values, beginning from the early design stages. The importance of such a system has recently been discussed in [32], stressing that its lack results in mostly "manual" approach to the evaluation of concurrency attributes satisfaction. The architecture of the system for routine design is presented in Figure 8.

In order to be concrete regarding the architecture proposed in Figure 8, its modules are considered next. Examples are provided to illustrate the application of the decision support techniques incorporated in the framework. While the examples are simple and hypothetical, they give the idea of the benefits to be expected from a fully developed system.

A. GROUPING OF PARTS IN THE DESIGN LIBRARY

Designer may deal with a large volume of information related to parts and subassemblies that have been designed previously. It is difficult to process all such information without decomposing the overall system into smaller, manageable subsystems. One way of grouping existing designs in the design library is according to their functions.

A typical design object can be decomposed into subassemblies and these in turn into parts. Design of each part is performed so that a certain functions are satisfied. Similar or identical functions may satisfy different parts. Figure 9 illustrates the grouping concept that is implemented in the proposed architecture.

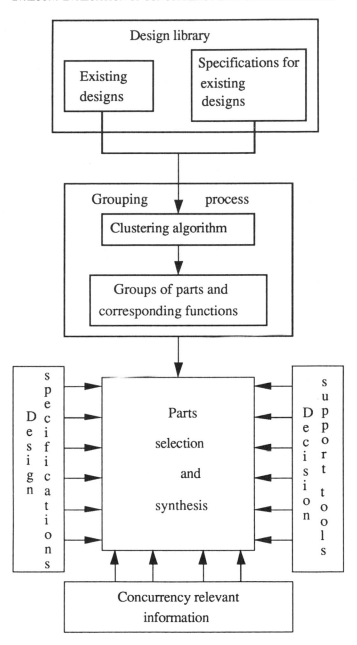

Figure 8. Architecture of the system for routine design

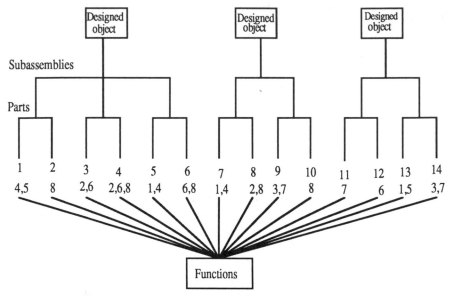

Figure 9. Designed objects and the corresponding functions

The hypothetical design library in Figure 9 consists of 14 parts and 8 functions. The correspondence between them can be represented as part-function incidence matrix $[a_{ij}]$:

Parts

$$
[a_{ij}] =
\begin{array}{c}
 & \begin{array}{cccccccccccccc} 1 & 2 & 3 & 4 & 5 & 6 & 7 & 8 & 9 & 10 & 11 & 12 & 13 & 14 \end{array} \\
\begin{array}{l} \text{F} \\ \text{u} \\ \text{n} \\ \text{c} \\ \text{t} \\ \text{i} \\ \text{o} \\ \text{n} \\ \text{s} \end{array}
\begin{array}{c} 1 \\ 2 \\ 3 \\ 4 \\ 5 \\ 6 \\ 7 \\ 8 \end{array}
&
\left[
\begin{array}{cccccccccccccc}
  &   &   &   & 1 &   & 1 &   &   &   &   &   & 1 &   \\
  &   & 1 & 1 &   &   &   & 1 &   &   &   &   &   &   \\
  &   &   &   &   &   &   &   & 1 &   &   &   &   & 1 \\
1 &   &   &   & 1 &   & 1 &   &   &   &   &   &   &   \\
1 &   &   &   &   &   &   &   &   &   &   &   & 1 &   \\
  &   & 1 & 1 &   & 1 &   &   &   &   &   & 1 &   &   \\
  &   &   &   &   &   &   &   & 1 &   & 1 &   &   & 1 \\
  & 1 &   & 1 &   & 1 &   & 1 &   & 1 &   &   &   &   \\
\end{array}
\right]
\end{array}
\qquad (16)
$$

where

$$
a_{ij} = \begin{cases} 1 & \text{if function } j \text{ is satisfied by part } i \\ 0 & \text{otherwise.} \end{cases}
$$

Transformation of the nonstructured matrix $[a_{ij}]$ in (16) into a structured one is a computationally complex problem. Cluster analysis provides an underlying theory for solving the grouping problem. Solving the problem represented in matrix (16) with the cluster identification algorithm presented in [33] produces matrix (17).

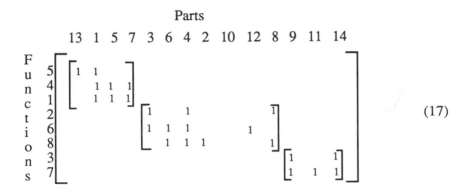

$$(17)$$

The decomposed matrix (17) results in the following:

(a) separation of the overall design information contained by the design library into groups of parts and functions,

(b) groupings of parts may differ from their traditional association with the design objects,

(c) given the required functions, all parts that satisfy these functions can be identified,

(d) complexity of management and retrieval of information from a design library is reduced.

B. DESIGN SPECIFICATIONS

Design specifications are formulated in an early design stage and belong to the most information-poor decision situations. However, the designer must choose among a number of alternative actions that may effect the degree of satisfaction of concurrency attributes. The designer's task is to make a rational use of concurrency related information available in order to select one of the design alternatives. The alternatives, at this stage of design, may, for example, involve selecting nodes in the AND/OR tree representing the space of requirements and functions [34]. A major difficulty associated with the application of techniques for handling uncertainty at this stage is the robustness of the prior-distribution caused by the scarcity of information. The minimax approach to handle the above problem is proposed. This approach was discussed in the decision-making context in Whalen [35] and in Lindley [36]. Blanchard and Fabrycky [23] proposed the minimax optimization as an evaluation tool in systems analysis. The minimax reasoning becomes more appropriate for the case with poor availability of information when the utility function is replaced with a *regret* function. Regret is the difference between the actual utility and the maximum possible value of utility. The minimax regret technique emerges from the fuzzy risk minimization for cases when the relative possibilities are totally unknown [35]. In practice, the minimax regret technique applied at early design stages directs a designer's attention to design paths which have the strongest effect on the degree of concurrency satisfaction.

A designer can enter the design library with clustered objects and functions after the specification stage has been completed. A formal approach to the design specification stage is presented in [34]. This approach takes into account two concurrency attributes, reliability and economic feasibility that are associated with design features such as cost and functions (see Figure 7 in this chapter) and the following generalized search algorithm is used:

The Generalized Search Algorithm
Step 1.   Form a one-element queue OPEN consisting of the root node.
Step 2.   If the first element of the queue OPEN is a goal node, stop;
                 Otherwise, expand the first element of the queue OPEN
generating   its child nodes and remove the first element from the queue
                 OPEN.
Step 3.   If only one child node is generated, then place the child node at the
                 beginning of the queue OPEN and go to Step 2; Otherwise, go
to   Step 4.
Step 4.   If there is qualitative knowledge associated with child nodes,

apply production rules to select nodes for further

consideration. The selected nodes are added to the queue

OPEN;

Otherwise, compute the cost of child nodes and add them to the
queue OPEN.

Step 5.   Sort the queue OPEN by estimated cost and go to Step 2.

Application of the above algorithm to the specification stage of a shaft
coupling is presented in [37]. Examples of the domain specific production
rules that are used in Step 4 of the algorithm for design of a shaft coupling are
presented next:

Rule 3

IF	a requirement is "Design a shaft coupling"
THEN	decompose it into "The nature of coupling is rigid"
AND	"Coupling is able to transmit torque"
OR	"The nature of coupling is flexible"
AND	"Coupling is able to transmit torque".

Rule 4

| IF | a requirement is "Coupling is able to transmit torque" |
| THEN | do not decompose it |

Rule 5

IF	a requirement is "The nature of coupling is flexible"
THEN	decompose it into "Coupling allows for radial offset of the shaft"
AND	"Coupling allows for axial offset of the shaft".

The minimax approach is incorporated into the generalized search algorithm. The "regret" function needed by the minimax technique is of qualitative nature and may either be attached to the node of the logic tree or may be selected by the designer. To illustrate the minimax approach to design decision situations the following example is used. Suppose that at a certain node of the logic tree a designer has to choose one of three nodes (N1, N2, and N3) connected by an OR connector. Economic feasibility is one of the design features that should be considered in concurrency analysis at an early design stage. At this design stage, frequently, no quantitative cost information is available and the minimax approach can be used to support qualitative data or judgement. This judgement is expressed as a hypothesis (H) about the level of regret in the degree of concurrency satisfaction taking into account the uncertainty of the final design outcome. Suppose the following is valid for the above case. If the designer chooses node N1, then $H_{1,1}$ may be experienced (no regret in the degree of concurrency satisfaction), $H_{1,2}$ (low regret), or $H_{1,3}$ (moderate regret), so the maximum possible regret for node N1 is $H_{1,3}$. If node N2 is chosen $H_{2,1}$ (high regret), $H_{2,2}$ (no regret), or $H_{2,3}$ (low regret), so the maximum possible regret for node N2 is $H_{2,1}$. Node N3 is associated with $H_{3,1}$ (very high regret), $H_{3,2}$ (high regret), or $H_{3,3}$ (no regret), so the maximum possible regret for N3 is $H_{3,1}$. Figure 10 presents the overall lattice of preferences and Figure 11 shows three sublattices for the three alternative nodes.

Note that the regret represented by $H_{1,2}$ is not necessarily equal to that represented by $H_{2,3}$; the same is true for $H_{3,2}$ and $H_{2,1}$. In the information-poor design decision situation, their magnitude relative to each other need not to be determined, only the magnitude relative to the rest of the lattice is to be considered. This makes the minimax approach a powerful tool to deal with uncertainty and incompleteness of information at the specification level. Another advantage of this approach is its flexibility. It can parallel the designer's experience and data availability by applying various regret levels and various assumptions about the final outcome of the design process. Besides, the minimax approach can be applied at the beginning of the design process to any concurrency attribute for which it is possible to define a corresponding regret .

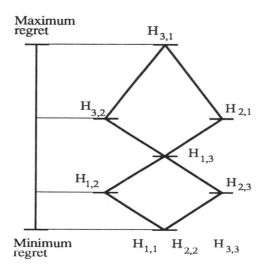

Figure 10. The overall lattice of regrets

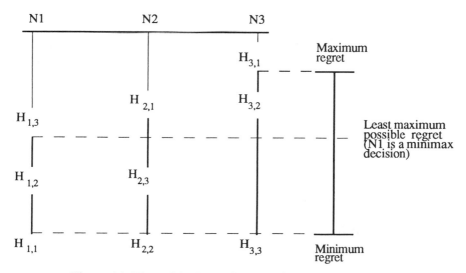

Figure 11. The sublattices of regret for alternative nodes

C. SELECTION OF PARTS

The set of the functions specified for a particular design determines the function-part clusters existing in the design library to be taken into further consideration. The decision problem at this phase of design is concerned with the selection of parts. Since in this design phase there is usually more concurrency relevant information available than in the specification phase, the approach based on the Dempster-Shafer theory [26] is proposed as a decision support tool for selection of parts. According to this approach the likelihood of a hypothesis H about the degree of concurrency satisfaction is represented as an interval $H[s(H),p(H)]$, where $s(H)$ is the minimum likelihood of H being true, and $p(H)$ is the maximum likelihood of H being true. The degree of uncertainty about the actual probability value of the hypothesis corresponds to the width of the interval $u(H)=p(H)-s(H)$. Hence, if $u(H)$ is zero then one deals with a point probability approach. To clarify the above, the following examples are considered:

$H[0,0]$ H is definitely false,

$H[1,1]$ H is definitely true,

$H[0.25,1]$ partial support is provided for H,

$H[0,0.85]$ partial support is provided for the negation of H,

$H[0.25,0.85]$ probability of H is between 0.25 and 0.85, i.e. the support

is   provided simultaneously for H and its negation.

The combined evaluation is computed according to the following rule:

Rule 1

IF $H_{1,1}[s_1(H_{1,1}),p_1(H_{1,1})]$

AND $H_{1,2}[s_2(H_{1,2}),p_2(H_{1,2})]$

THEN $H1[s(H_1)p(H_1)]$

where $s(H1)=\max[s_1(H_{1,1}),s_2(H_{1,2})]$ and $p(H_1)=\min[p_1(H_{1,1}),p_2(H_{1,2})]$. To illustrate the preceding, the following example is considered. Assume that among the functions selected at the specification phase there are functions F2 and F6 that belong to the following cluster (matrix (17)):

$$
\begin{array}{c}
\quad\;\; P3\;\; P6\; P4\; P2\; P10\; P12\; P8 \\
\begin{array}{c} F2 \\ F6 \\ F8 \end{array}
\left[
\begin{array}{ccccccc}
1 & & 1 & & & & 1 \\
1 & 1 & 1 & & & 1 & \\
& 1 & 1 & 1 & 1 & & 1
\end{array}
\right]
\end{array}
\qquad (18)
$$

As shown in matrix (18), two parts (P3 and P4) satisfy functions F2 and F6. Assume that based on the data available, or judgement of the designer, the minimum (s) and the maximum (p) likelihood of the degree of satisfaction of the maintainability attribute (HM) is given in Table I.

Table I. Minimum and maximum likelihood of the degree of satisfaction of maintainability

Function-part correspondence	Minimum likelihood (s)	Maximum likelihood (p)
F2,P3	0.25	0.85
F6,P3	0.25	1
F2,P4	0.5	0.85
F6,P4	0.5	1
F8,P4	0.25	1

According to the data in Table I, the following likelihoods are computed for parts P3 and P4 (Rule 1);

$$P3 \quad HM[0.25, 0.85]$$

(19)

$$P4 \quad HM[0.5, 0.85]$$

As indicated in Eq. (19), the degree of uncertainty for part P4 is smaller than for part P3 and the decision is to choose part P4. A similar procedure is applied for all clusters that include functions resulting from the parts selected. The parts selected are further synthesized into the designed object.

The Dempster-Shafer approach enables the designer to deal with various information sources (for example, opinions of different experts). It is assumed that a knowledge source KS_1 distributes a unit of belief across a set of concurrency attributes for which a direct evidence for support or negation exists. Thus a set of attributes, say $H_{1,1}$, $H_{1,2}$, $H_{1,3}$, and so on has a total sum of belief equal to 1, i.e., $\sum m_1(H_{1,i}) = 1$, where $m_1(H_{1,i})$ represents the portion of belief that KS_1 has committed to the hypothesis of attribute H_1, called the basic probability mass. If the second knowledge source KS_2 is applied (or the opinion of another expert is taken into consideration) the information is combined by computing the orthogonal sum as illustrated with the hatched rectangle in Figure 12.

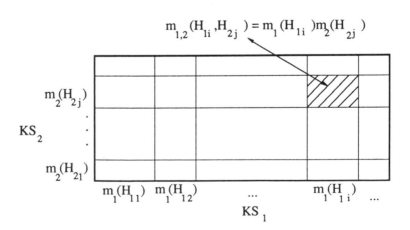

Figure 12. Composition of the basic probability mass from knowledge sources KS_1 and KS_2

Figure 12 shows the vertical strip of measure $m_1(H_{1,i})$ commited to $H_{1,i}$ by KS_1, and the horizontal strip of size $m_2(H_{2,j})$ commited to $H_{2,j}$ by KS_2. The intersection of these strips commits $m_1(H_{1,i})m_2(H_{2,j})$ to the combination of $H_{1,i}$ and $H_{2,j}$.

The Dempster-Shaffer methodology has a number of advantages that are important from the designer's point of view:

(i)   The precision of a hypothesis about concurrency satisfaction is clearly indicated by the difference $p(H_1)-s(H_1)$. If the difference is small, then the knowledge about H_1 is relatively precise, if it is large then correspondingly little is known. If $p(H_1)=s(H_1)$, then the knowledge of H_1 is exact and the point probability approach can be used.

(ii)   The amount of support for and against the degree of concurrency satisfaction is easily understood. The amount of support for H_1 is at least $s(H_1)$ and the amount of support for the negation of H_1 is $1-p(H_1)$.

(iii)   The methodology is developed on the basis of the probability theory where $s(H_1)$ and $p(H_1)$ are interpreted.

D. SYNTHESIS OF PARTS

The rule-based approach to design synthesis is presented in [39]. This approach was implemented in an object-oriented programming environment (Smalltalk-80) in the Intelligent Systems Laboratory at the University of Iowa [40]. The implementation is based on the following synthesis algorithm:

Algorithm
Step 1.   Open the set MODEL_BASE consisting of all design building blocks. Set level = 1.

Step 2.   Apply production rules to generate connections between elements in MODEL_BASE.

Step 3.   If no connections are generated, stop; Otherwise, match elements in MODEL_BASE into pairs using the existing connections.

Step 4.   Define input and output variables for models generated by the matching process.

Step 5.   Remove from MODEL_BASE all elements that have taken part in the matching process.

Step 6.   Add to MODEL_BASE all models generated by the matching process.

Step 7.   Set level = level + 1 and go to Step 2.

Two examples of production rules used in Step 2 are presented next.

Rule 7

IF   two elements in the model base have identical output and input variables

AND   there are no production rules that prevent their connection

THEN specify the connections for these elements

Rule 8

IF   there are no elements in the model base with identical input and output variables

AND   there are elements with partially identical input and output variables

AND   there are no production rules that prevent their connection

THEN specify a connection for these elements beginning with the closest match

The object-oriented programming environment has been chosen for implementation of the synthesis algorithm becouse the basic characteristics of the object-oriented paradigm: *abstraction, encapsulation, inheritance, and polymorphism*, ideally support the process of design synthesis. *Design abstraction*, in the sense usually employed at the conceptual design level, is a representation for a complicated design object or its fragments. Details of such abstraction are not essential to understanding of its purpose and functionality.

The abstraction is provided by creating objects with a set of characteristics and messages assigned to them. For example, in order to represent a component to be used in design synthesis, a class **Component** (a class is also an object) representing that entity can be created. A set of characteristics typical for a particular component has to be defined as well as a set of messages on which an instance of the class **Component** will respond.

Components that take part in the design synthesis, are *encapsulations* of abstractions. Encapsulation represents a process of implementing data abstraction by defining objects that have properties which are either shared with other objects of the same class or are private. Instance variables and methods defined in a class description protocol represent encapsulation means. For example, some instance variables in the class **Component** include component identification, component input and output, and so on. All instances of this class inherit the same instance variables, but with different values. Instances of a class respond to the set of messages defined in the class description protocol. For example, if there are eight components to be considered in design synthesis, all instances of the class **Component** will respond to the message getComponentName in the same way, i.e. by returning the identification of a particular component.

Two components may be matched into a model and two models into another model at the higher level of hierarchy. Lower level models in such hierarchy share at least some features of the upper level ones. Models of the lower level can be described as more specialized elements of the overall model. It means that they have at least the same properties as models of the higher level. Due to their more specialized nature, lower level models may have new properties that are not descriptive of the more general models (they inherit the properties of the higher level models). This is the analogy to what in object-oriented programming is called *inheritance*. Inheritance can be realized by defining an object that is a subclass of an existing class. This object inherits properties defined in its superclass. For example, the class **Component** is defined as a subclass of the class **ComplexModel**, and it inherits information defined in the class description protocol defined in its superclass. The hierarchical structure of the synthesized model parallels the inheritance hierarchy of objects in object-oriented programming.

Input variables of components are defined as flexible sensors that can accept values of various attributes. It is possible to input the same value to different components and have each component respond in a way appropriate to its defined behavior. The above supports flexibility of the matching process and is called *polymorphism*.

Thus, it is clearly seen that the synthesis methodology can be characterized by the following four component and model phenomena: abstraction, encapsulation, inheritance, and polymorphism. This suggested that it could be supported by object-oriented programming languages that provide these characteristics.

The object-oriented synthesis is included in the presented system as one of its components. A mechanism for monitoring and evaluation of the degree of concurrency satisfaction is incorporated into the system as one of its features. The idea of monitoring and updating of the degree of concurrency satisfaction is illustrated in Figure 13.

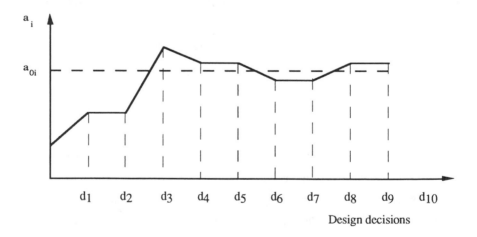

Figure 13. Monitoring the $a_i \dashrightarrow a_{0i}$ correspondence at various decision points during the design synthesis

In Figure 13, a_i denotes the actual degree of satisfaction of ith concurrency attribute at each synthesis step, and a_{0i} denotes the required degree (a_i and a_{0i}

correspond to the notation used in Figure 5). The decision points, represented in Figure 13 as d_i, involve adding previously selected parts to the designed object. Such an updating process is in part related to the accuracy of information concerning contribution of various parts to the degree of satisfaction of concurrency attributes. Thus the approach must take intoaccount the uncertainty aspect. The Bayesian approach is a tool for such a formulation that improves the estimates whenever an additional part is added (Hines and Montgomery [41]). This approach is suitable for an information-rich environment at the synthesis stage, as here there might be enough information to switch from the "two-valued" Dempster-Shafer approach into the point ("one-valued") probability approach. Bayes' theory was successfully incorporated into the development of inference engines (see, for example, Duda *et al.* [42] and Forsyth [43]) and proved to be useful in dealing with some ill-defined problems [44]. The Bayes approach helps to combine sample information about concurrency with the prior information in evaluation of the degree of concurrency satisfaction during the conceptual design process. The probability associated with the prior information is called a *subjective probability*, in that it measures a designer's level of belief in a given proposition. The designer uses his own experience and knowledge as the basis for arriving at a subjective probability. Given a priori probability about the initial degree of satisfaction of a certain concurrency attribute, the designer may modify this probability to produce a posteriori probability while moving through the design steps and adding some new components. The process of updating is repeated and each time the probability related to the degree of concurrency satisfaction is shifted up or down using a Bayesian rule with a different prior probability derived from the last posterior probability. In the end, having collected all the information available, the designer can reach a final conclusion using upper and lower thresholds, defining the areas of acceptance and rejection. Such thresholds are set according to the importance of a given concurrency attribute. In Figure 14, the areas of acceptance and rejection are shown for an attribute reliability of a coupling.

For the design in Figure 14, reliability is considered to be an important attribute because it has been assigned a narrow area of acceptance and wide area of rejection.

The Bayesian theory seems to be a suitable approach that resembles the designer's way of reasoning. It proceeds under the assumption that the designer has a certain prior judgement about the degree of the concurrency

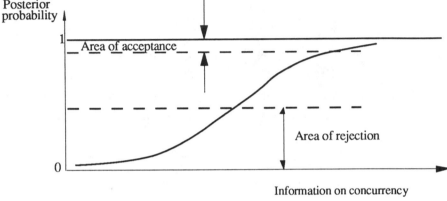

Figure 14. Upper and lower thresholds of the posterior probability of the degree of satisfaction of an attribute "reliability"

satisfaction and that this judgement can be expressed numerically. When a designer adds a new part to the designed object, the Bayes' rule allows to represent the posterior probability of the new state of satisfaction. The following example is used to provide the interpretation of the Bayesian approach and its relation to the design process. Suppose that at a certain point of design synthesis the probability of satisfaction of attributes "safety" and "manufacturability" is estimated by the designer as 0.5. The designer adds the part P4 that would posses a probability of 0.8 of satisfying a manufacturability attribute, and 0.6 of satisfying a safety attribute. The question that arises is: when the designer adds part P4, what is the probability of the above attributes actually being satisfied? Applying the Bayes' rule in Eq. (20)

$$P(B_k)P(A \mid B_k)$$

$$P(B_k|A) = \frac{}{\sum P(B_i)P(A \mid B_i)} \qquad \text{for i= 1,2, ...,n} \qquad (20)$$

to the manufacturability attribute one obtains:

$$P(B) = 0.5, P(NOT\ B) = 0.5, P(A \mid B) = 0.8, P(A \mid NOT\ B) = 0.2. \quad (21)$$
Thus:

$$P(B \mid A) = \frac{P(B)P(A \mid B)}{P(B)P(A \mid B) + P(NOT\ B)P(A \mid NOT\ B)} = 0.8. \qquad (22)$$

For the safety attribute we have:

$$P(B) = 0.5, P(NOT\ B) = 0.5, P(A \mid B) = 0.6, P(A \mid NOT\ B) = 0.4, \quad (23)$$

which gives:

$$P(B \mid A) = 0.6. \qquad (24)$$

Thus, this particular design step has increased the prior probability of the attribute manufacturability being satisfied from 0.5 to a posterior probability of 0.8. At the same time, it has increased the prior probability of the attribute safety being satisfied from 0.5 to to a posterior probability of 0.6. From this point, the designer assumes 0.8 and 0.6 as being new a priori probabilities for the attributes manufacturability and safety respectively.

For the overall design (after the synthesis has been completed), the final probabilities can be checked against the lower and upper thresholds of acceptance for each concurrency attribute and thus providing a tool for acceptance of the design from the concurrency satisfaction point of view. The Bayesian approach represents a flexible technique which keeps the designer involved in the concurrency evaluation and thus, enables him to learn and benefit from the design process itself. Other advantages of the Bayesian technique of handling uncertainty and incomplete information concerning the satisfaction of concurrency attributes are as follows:

- it is based on the probability theory which is easy to understand and which seems to mirror the natural way of thinking,

- it represents straightforward computational procedure that can be readily implemented,

- final posterior probabilities are not affected by the order in which the components are connected.

One of the most important elements of making a design decision is the early choice of an appropriate formal model. Bearing the above in mind, three different approaches are proposed. The goal is to maximize the efficient use of whatever information is actually available to a designer while minimizing the need for introducing arbitrary assumptions. Such assumptions are inevitable when only one technique is at a disposal and this is usually the case in research reported in the literature. For example, suppose that the information available in a given situation is sufficient for the Dempster-Shafer approach. Then, to use a less information-rich technique such as the minimax regret would ignore real information that might be critical to the evaluation process. On the other hand, to use a more information-rich approach would always require introduction of arbitrary assumptions that might lead to a undesirable decision.

E. TRANSFORMATION FROM CONCEPTUAL TO EMBODIMENT DESIGN

Embodiment design usually begins with a rough layout derived from the concept of the designed object [11, 45]. This means that one needs a transformation from an abstract concept to the concrete layout. In this layout there are usually clusters of components (assemblies) that must be given a special consideration as far as their geometry and other physical attributes are concerned. For example, as presented in Pahl and Beitz [11] a shaft and a hub may be treated as components of such a cluster as the design of their geometry determines the deflection of a force flowline. The same applies to such components as hub and key, or hub, key, and shaft (Orthwein [46]). One may say that components in a cluster are strongly connected in a sense that their design at the embodiment stage is strongly interdependent (can not be pursued separately). The connectedness has also further implications on process planning, machining, assembly, and so on. A strong connection between components within a cluster is determined by the flow of energy, material, and/or signals (Pahl and Beitz [11]), and its discovery at an early stage of the design process may be useful in transformation from the conceptual to embodiment design. So far, no formal approach to such transformation exists. On the other hand, the problem itself is an important one and its solution has been recognized as a necessary step in the development of an intelligent computer aided design system [47]. In the presented system the above transformation problem is solved using some concepts from graph theory. The following transformation algorithm is applied (Kusiak and Szczerbicki [48]).

Transformation Algorithm

Step 0. Develop digraph G of the flow of energy between components of the designed object.

Step 1. Define an nxn adjacency matrix A of digraph G, where n equals the number of vertices.

Step 2. Calculate matrix $E = \sum\limits_{r=1}^{n-1} A^r$, where A^r is equal to A to the power of r.

Step 3. Add unity matrix I to E.

Step 4. Calculate reachability matrix $R = B(I + E)$, where B denotes Boolean function.

Step 5. Calculate the product $R \cdot R^T$.

Step 6. Apply cluster identification algorithm to the matrix $R \cdot R^T$.

Step 7. Develop condensation of the digraph G.

Next, the design example of a holding device is shortly discussed to illustrate the application of the above algorithm. The synthesis stage of conceptual design of a holding device has been presented in [40]. For this design the set of design functions is given by {F1, F2, F3, F4, F5, F6, F7, F8, F9, F10, F11}, where:

F1:   apply human force rotationally and amplify the force,

F2:   transform the force by one screw,

F3:   hold material by a movable part of the device,

F4:   hold material by a fixed part of the device,

F5:   pass force to a fixed holding part of the device,

F6:   keep material between movable and fixed parts of the device,

F7:   guide the motion of a movable part by the fixed guide,

F8:   fix the unmovable part of the device,

F9:   pass force and velocity to the movable part of the device,

F10:   pass the holding force to the guide of a movable part,

F11:   fix the guide of a movable part.

The above has been obtained from the design requirements using the optimization procedure discussed in [34]. The set of functions is treated in transformation algorithm as the basis for the digraph representation of the holding device. Following the logic of the functions and requirements, and using information that they contain, the vertices of the digraph corresponding to functions F1 through F11 are defined as follows:

Vertex 1.   Point at which the energy (human force) enters the device.

Vertex 2.   Point at which the energy is (after entering the device) transformed.

Vertex 3.   Point at which a movable holding part exercises holding
force   on the material.

Vertex 4.   Point at which a fixed holding part exercises holding force
on   the material.

Vertex 5.   Point at which energy enters a fixed holding part of the
device.

Vertex 6.   Point at which holding forces are passed from fixed and movable parts to the material that is being held.

Vertex 7.   Point at each fixing force interacts with the guide.

Vertex 8.   Point at which holding force is compensated by the fixing force in the unmovable part of the device.

Vertex 9.   Point at which the transformed energy enters the movable part   of the device.

Vertex 10.   Point at which the holding force interacts with the guide of a movable part.

Vertex 11. Point at which the guide of movable part is fixed by fixing force.

The above vertices define the digraph G of the rough flow of energy in the device as presented in Figure 15.

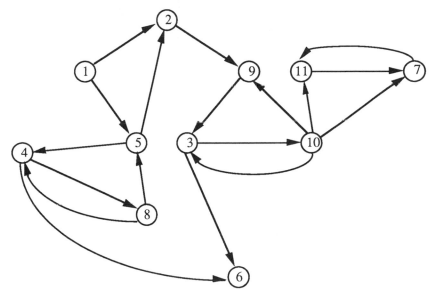

Figure 15. Digraph G of the flow of energy in the holding device

Applying the transformation algorithm to the digraph G in Figure 15 one arrives at the condensation as presented in Figure 16.

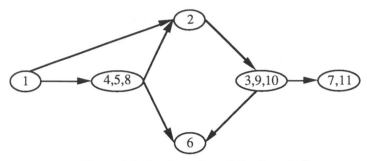

Figure 16. Condensation of the digraph G

Cluster C-1 = {1} includes function F1. Cluster C-2 = {4, 5, 8} functions F4, F5, and F8. Functions F3, F9, and F10 are included in cluster C-3 = {3, 9, 10}. Finally, cluster C-4 = {7, 11} includes functions F7 and F11. The condensation suggests the physical layout of the design as it defines the way the clusters should be connected. As suggested by the clusters generated, the layout of the holding device that corresponds to the condensation in Figure 16 may include a lever, a holding part that is fixed, a screw, a holding part that is movable, and a guiding element. Also, the material that is held by the device plays important role in the layout (Figure 17).

The presented algorithm was implemented in an object-oriented programming language C++ [49]. A number of promising applications of the object-oriented approach in various areas of engineering have been reported in the literature. Birth and Whiteley [50] developed a real-time control expert system using the Smalltalk-80 programming environment. Keirouz *et al.* [51] presented an object-oriented domain model for constructed facilities. Regourd [52] proposed application of object-oriented programming for an intelligent engineering tutoring system. Fenves [53] described application of Smalltalk for engineering software development. Powell and Bhateja [54] used objects for development of engineering data bases. Fenves [55] applied an object-oriented approach for structural design and analysis. In [40] an object-oriented system for conceptual design was proposed. Also, the development of an intelligent CAD system with object-oriented environment was suggested in [56].

The architecture of an object-oriented software system is built using a set of

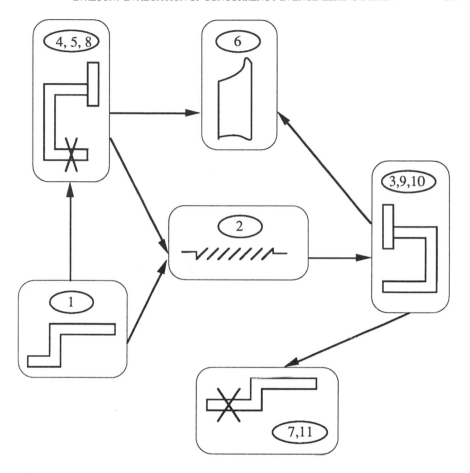

Figure 17. Physical layout corresponding to the condensation of the digraph
    G shown in Figure 16

classes that characterize the bahavior of all data involved. The behavior of
each class is further defined by corresponding methods. An object-oriented
implementation of the transformation algorithm presented in this chapter was
suggested by the presence of a number of matrix manipulations involved.
Thus, a matrix is treated as an object and a matrix class is built for
implementation of addition, subtraction, multiplication, and transpose
functions. All these manipulations are coded as methods and can be sent to the
class **Matrix** on which they are executed. In this way, the object-oriented
code is created that naturally represents the problem space and easily maps

into entities in this space. For the retrieval of clusters in a given digraph, the class **Cluster** is developed that inherits all methods from the **Matrix** class. In addition it provides new data (list of clusters) and methods (draw_horizontal_line, draw_vertical_line) for implementation of the cluster identification algorithm. It also calls the Domain Graphics Primitive Resource Program to produce a graphical output.

IV. CONCLUSIONS

Design of products in a concurrent engineering environment is a critical process. It is viewed as a strategic task which has a major effect on subsequent manufacturing and service activities. For example, design of some mechanical systems determines 70% to 80% of the final production cost [57]. Another study included in [24] showed that 70% of the life cycle cost of a product is determined at the conceptual design stage. Thus, tools are needed to enable the designer to perform the evaluation of concurrency attributes of the designed product as the design evolves. One can clearly see that the concepts presented in this chapter increase the level of concurrency satisfaction of various concurrency attributes.

Evidential reasoning, which is the task of inferring the likelihood of some hypotheses about the level of concurrency satisfaction, is central to many computer systems that help users in decision making and diagnosis. The problem of evidential reasoning is complicated by the fact that information being conveyed by a piece of evidence is often not only uncertain, but also imprecise, incomplete, and vague. A large part of the research published on evidential reasoning is based on the following theoretical frameworks: mini-max optimization, Bayesian probability theory, Dempster-Shafer theory, and fuzzy set theory. These frameworks differ in their strengths and weaknesses. For example, the Bayesian probability theory has a well-developed decision-making mechanism, but it requires precise probability judgments. The Dempster-Shafer is based on the probability theory, yet it allows probability judgements to capture the imprecise nature of the evidence. As a result, the degree of likelihood is measured with probability intervals, as opposed to the point probability in the Bayesian approach. One of the weaknesses of the Dempster-Shafer theory is that it is still not well developed. The fuzzy set theory focuses on the issue of representing and managing vague information. One of its strengths is its possibility theory as a foundation for dealing with imprecise data. Application of the fuzzy set theory is considered to be a long-term goal in further development of the system presented in this chapter.

The approach presented advocates that the decision making models play a significant role in helping designers to incorporate and evaluate concurrency requirements at various stages of the product development. The system assists a designer to assess a candidate design with respect to various concurrency attributes. It is effective not only as a preliminary screening for various life-cycle values but for the designer training as well. The designer working with the system is continuously involved in concurrency evaluation and s/he benefits from this involvement.

The system presented improves product quality and shortens the product development time. It ensures that concurrency attributes are incorporated into the design before its completion and they are satisfied to a degree required. Failure to do so may lead to a costly redesign and lengthy development cycle.

The system is embedded in an existing library of objects, parts, and corresponding functional specifications. It assists a designer in selecting of existing parts and synthesizing them into a new design. The system:

(a)   can be used interactively by the designer,

(b)   groups parts and functions,

(c)   allows for incremental analysis of the design as it evolves,

(d)   assists a designer in evaluation of the design from different concurrency perspectives,

(e)   deals with uncertainty and incompleteness of information at various design stages,

(f)   provides a comprehensive set of tools suitable for different levels of uncertainty and availability of information.

The long-term research goals associated with the presented system are as follows:

• Expansion of the library of evaluation tools by including other approaches capable of handling uncertainty. In particular Z-fuzzy and L-fuzzy based techniques will be considered to deal numerically with the linguistic hedges that frequently appear in evaluations provided by design experts. Such hedges include, for example, "very", "much", "approximately", and "less". Morphological analysis will also be considered as an alternative way of finding the best combination of the existing parts and functions.

• The extension of the system that makes it more flexibile is planned. This flexibility will be achieved by allowing a designer to decide which approach to use independently of the actual design stage. Such decision will be based exclusively on the availability of concurrency related data. The tools available will not be tied to a particular design activity.

References

1. G. Nadler, "Systems Methodology and Design," *IEEE Transactions on Systems, Man, and Cybernetics,* **15**, 685-697 (1985).
2. W. Asimow, "Introduction to Design," Prentice-Hall, Englewood Cliffs, N.J., 1962.
3. R.D. Coyne, M.A. Rosenman, A.D. Radford, M. Balachandran, and J.S. Gero "Knowledge-based Design Systems," Addison-Wesley, Reading, MA., 1990.
4. B.L. Archer, "The Structure of the Design Process," *in* "Design Methods in Architecture," (G. Broadbent and A. Ward, eds), Lund Humphries, London, 1969.
5. F. Mistree and D. Muster, "Conceptual Models for Decision-Based Concurrent Engineering Design for the Life Cycle," *Proceedings of the Second National Symposium on Concurrent Engineering,* Morgantown, W.V., 443-468 (1990).
6. M.D. Mesarovic, D. Macko, and Y. Takahara, "The Theory of Hierarchical Multilevel Systems", Academic Press, New York, 1970.
7. A.D. Radford and J.S. Gero, "Design by Optimization in Architecture, Building, and Construction," Van Nostrand Reinhold, New York, 1988.

8.   S. Arora, "Introduction to Optimum Design," McGraw-Hill, New York, 1989.

9.   C.J. Koomen, "The Entropy of Design: a Study on the Meaning of Creativity," *IEEE Transactions on Systems, Man, and Cybernetics,* **15,** 16-30 (1985).

10. R.C. Conant, "Laws of Information which Govern Systems," *IEEE Transactions on Systems, Man, and Cybernetics,* **6,** 240-255 (1976).

11. G. Pahl and W. Beitz, "Engineering Design," Springer-Verlag, New York, 1988.

12. V. Hubka, "Principles of Engineering Design," Butterworth Scientific, London, UK., 1982.

13. E. Simoudis, "A Knowledge-Based System for the Evaluation and Redesign of Digital Circuit Networks," *IEEE Transactions on Systems, Man, and Cybernetics,* **8,** 302-315 (1989).

14. L.I. Steinberg and T.M. Mitchell, "A Knowledge Based Approach to VLSI CAD - the REDESIGN System," *Proceedings ACM IEEE 21st Design Automation Conference,* Albuquerque, N.M., 412-418 (1984).

15. R. Ellsworth, A. Parkinson, and F. Cain, "The Complementary Roles of Knowledge-Based Systems and Numerical Optimization in Engineering Design Software," *ASME Journal of Mechanisms, Transmissions, and Automation in Design,* **111,** 100-103 (1989).

16. U. Flemming, R. Coyne, T. Glavin, and M. Rychener, "A Generative Expert System for the Design of Building Layouts - Version 2," *in* "Artificial Intelligence in Engineering Design," (J.S. Gero, ed.), Elsevier, New York, 1988.

17. J.S. Gero and R.D. Coyne, "Knowledge-Based Planning as a Design Paradigm," *in* "Design Theory for CAD," (H. Yoshikawa and E.A. Warman, eds), Elsevier, New York, 1987.

18. M.L. Maher, "HI-RISE: an Expert System for Preliminary Structural Design," *in* "Expert Systems for Engineering Design", (M.D. Rychener, ed.), Academic Press, Boston, 1988.

19. D.E. Grierson and G.E. Cameron, "An Expert System for Structural Steel Design," *in* "Artificial Intelligence in Engineering: Design," (J.S.

Gero, ed.), Elsevier, New York, 1988.

20. S.N.T. Shen, M-S. Chew, and G.F. Issa, "Expert System Approach for Generating and Evaluating Engine Design Alternatives," *in* "Applications of Artificial Intelligence," (M.M. Trivedi, ed.), The International Society of Optical Engineering, Bellingham, Washington, 1989.

21. H.A. Simon, "The Sciences of the Artificial," The MIT Press, Cambridge, M.A., 1982.

22. A. Rapoport, "General Systems Theory," Abacus Press, Turnbridge Wells, Kent, U.K., 1986.

23. B.S. Blanchard and W.J. Fabrycky, "Systems Engineering and Analysis," Prentice Hall, Englewood Cliffs, N.J., 1990.

24. J.L. Nevins and D.E. Whitney, "Concurrent Design of Product and Processes," McGraw-Hill Publishing Company, New York, 1989.

25. M.S. Mesarovic and Y. Takahara, "Abstract Systems Theory," Springer-Verlag, New York, 1989.

26. K. Ishii, A. Goel, and R.E. Adler, "A Model of Simultaneous Engineering Design," *in* "Artificial Intelligence in Design," (J.S. Gero, ed.), Springer-Verlag, New York, 1989.

27. M.L. Maher, "Synthesis and Evaluation of Preliminary Designs," *in* "Artificial Intelligence in Design," (J.S. Gero, ed.), Springer-Verlag, New York, 1989.

28. K.L. Wood, K.N. Otto, and E.K. Antonsson, "A Formal Method for Representing Uncertainties in Engineering Design," *Proceedings of the First International Workshop on Formal Methods in Engineering Design, Manufacturing, and Assembly*, Colorado Springs, Colorado, 202-246 (1990).

29. S. Kota, A.G. Erdman, D.R. Riley, A. Esterline, and J.R. Slagle "A Network Based Expert System for Intelligent Design of Mechanisms," *AI EDAM,* **2**, 17-32 (1988).

30. W. Zhang and J. N. Siddall , "A Model for Intuitive Reasoning in Expert Systems," *ASME Computers in Engineering,* **1**, 459-466 (1988).

31. D.J. Wilde, "Globally Optimal Design," Wiley, New York, 1978.

32. T. L. De Fazio , A.C. Edsall, R.E. Gustavson, J. Hernandez, P. M. Hutchins, H.-W. Leung, S. C. Luby, R. W. Metzinger, J. L. Nevins, K. Tung, and D. E. Whitney, "A Prototype for Feature-Based Design for Assembly," *Proceedings of the Second National Symposium on Concurrent Engineering,* February 7-9, 1990, Morgantown, W.V., 559-593 (1990).

33. A. Kusiak, "Intelligent Manufacturing Systems," Prentice Hall, Englewood Cliffs, N.J., 1990.

34. A. Kusiak and E. Szczerbicki, "A Formal Approach to Design Specifications," *Proceedings of the ASME Design Automation Conference,* Chicago, IL, 311-316 (1990).

35. T. Whalen, "Decisionmaking Under Uncertainty with Various Assumptions about Available Information," *IEEE Transactions on Systems, Man, and Cybernetics,* **14**, 888-900 (1984).

36. D. V. Lindley, "Making Decisions," Wiley, New York, 1974.

37. A. Kusiak, E. Szczerbicki, and K. Park, "A Novel Approach to Decomposition of Design Specifications and Search for Solutions," *International Journal of Production Research,* (1991), (in press).

38. J.A. Garvey, J.D. Lowrance , and M.A. Fischler, "An Inference Technique for Integrating Knowledge from Disperate Sources," *Proceedings of the 7th International Joint Conference on AI,* Vancouver, B.C., pp. 51-62 (1981).

39. A. Kusiak and E. Szczerbicki, "Rule-Based Synthesis in Conceptual Design," *Preceedings of the Third International Symposium on Robotics and Manufacturing: Research, Education, and Applications (ISRAM'90),* **3**, Vancouver, B.C., 757-762 (1990).

40. A. Kusiak, E. Szczerbicki, and R. Vujosevic, "Intelligent Design Synthesis: an Object Oriented Approach," *International Journal of Production Research,* (1991), (in press).

41. W. W. Hines and D. C. Montgomery, "Probability and Statistics in Engineering and Management Science," Wiley, New York, 1980.

42. R.O. Duda, P.E. Hart, and N.J. Nilsson, "Subjective Bayesian

Methods for Rule-Based Inference System," *Readings in Artificial Intelligence,* Tioga, CA, 112-129 (1981).

43. R. Forsyth, "Expert Systems - Principles and Case Studies," Chapman and Hall, London, UK., 1984.

44. E. Szczerbicki, "Bayesian Approach to Incomplete Information in Group Functioning," *Progress in Cybernetics,* **10**, 115-122 (1987).

45. K. Roth, "Design Models and Design Catalogs," *Proceedings of the International Conference on Engineering Design,* Boston, Mass., 1987.

46. W. Orthwein, "Machine Component Design," West Publishing Company, St. Paul, M.N., 1990.

47. F. Kimura, "Architecture of Intelligent CAD System," *in* "Intelligent CAD, I ," (H. Yoshikawa and D. Gossard, eds), North-Holland, Amsterdam, 1989.

48. A. Kusiak and E. Szczerbicki, "A Methodology for Transformation from Conceptual to Embodiment Design, Working Paper No. 90-14, University of Iowa, Department of Industrial Engineering, IA (1990).

49. S. B. Lippman, "C++ Primer", Addison-Wesley, Mass., 1989.

50. M. Birch, and K. Whiteley, "An Object-Oriented Expert System Based on Pattern Recognition," *IEEE Transactions on Systems, Man, and Cybernetics,* **20**, 33-44 (1990).

51. W. T. Keirouz, D. R. Rehak, and I. J. Oppenheim, "Development of an Object Oriented Domain Model for Constructed Facilities," *in* "Artificial Intelligence in Engineering," (D. Sriram and R.A. Adey, eds), Computational Mechanics Publications, Old Woking, Surrey, UK, 1987.

52. J-P. Regourd, "Objects and Inferences: Another Approach for Intelligent Tutoring Systems," *in* "Artificial Intelligence in Engineering," (D. Sriram and R.A. Adey, eds), Computational Mechanics Publications, Old Woking, Surrey, UK, 1987.

53. G. L. Fenves, "Object Oriented Programming for Engineering Software Development," *Engineering with Computers,* **6,** 6-15 (1990).

54. G. H. Powell and R. Bhateja, "Data Base Design for Computer-Integrated Structural Engineering," *Engineering with Computers,* **4,**

135-143 (1988).

55. G. L. Fenves, "Object Representations for Structural Analysis and Design," *Proceedings of Fifth Conference on Computing in Civil Engineering,* ASCE, Alexandria, VA, 502-511 (1988).

56. T. Tomiyama , T. Kiriyama , H. Takeda,   D. Xue, and H. Yoshikawa, "Metamodel: A Key to Intelligent CAD Systems," *Research in Engineering Design,* **1,** 19-34 (1989).

57. J. Corbett, "Design for Economic Manufacture," *Annals of CIRP,* **35,** 93 (1986).

CONCURRENT ENGINEERING: DESIGN OF ASSEMBLIES FOR SCHEDULABILITY

ANDREW KUSIAK

Intelligent Systems Laboratory
Department of Industrial Engineering
The University of Iowa
Iowa City, IA 52242

I. INTRODUCTION

The traditional design of assemblies (products) has relied on an iterative approach. The main difficulty with the iterative approach is that it is time consuming and many iterations are required before a design project is completed.

In order to reduce the design cycle, concurrent design has been introduced. The basic idea of concurrent design is to shorten the time horizon in which the design constraints, such as schedulability, manufacturability, quality, reliability, maintainability, and so on are introduced. This paper focuses on the schedulability constraint. It will be shown that assembly design has a significant impact on the complexity of the scheduling problem and its solution quality. Two significant aspects of assembly design are considered:

(1) the structure of precedence constraints, and (2) the number of assembly plans.

The literature on the link between design of assemblies and operations of manufacturing systems is scarce. However, a number of papers on scheduling assembly systems have been published. A comprehensive review and analysis of the assembly line balancing problem was reported in Ghosh and Gagnon [4]. The numerous quantitative and qualitative factors mentioned in the literature that could impact the design, balancing, and scheduling of assembly systems are organized into an eight-level hierarchical taxonomy. Udomkesmalee and Dagnanzo [7] focused on unpredictable job sequences which result from parallel processing of job sequences in a flexible assembly system. They provided strategies for dealing with jobs getting out of sequence. Donath and Graves [2] presented an algorithm (SCHEDULE) for the near real-time dispatching and routing of multiple products in a flexible assembly system (FAS). They tested the algorithm for large and complex assemblies, large workloads, parametrization on the urgency factor setting in SCHEDULE, randomization of job due dates, and the distribution of the job release dates for the workloads assigned to the FAS. Wilhelm and Shin [6] described a study which investigated the influence that alternate operations might have on the performance of a flexible manufacturing system. Results showed that alternate operations reduce flow (cycle) time while increasing machine utilization. Fry *et al.* [3] studied the performance of fourteen sequencing rules in a six-machine assembly job shop. Their results indicated that there is a strong relationship between product structure and the performance of sequencing rules.

In section II of this chapter, the assembly scheduling problem is introduced. Two variants of scheduling problem and the corresponding algorithms are discussed in section II and IV: (1) fixed-assembly plan problem, and (2) N-assembly plan problem. Computational results and conclusions are discussed in sections V and VI, respectively.

II. THE ASSEMBLY SCHEDULING PROBLEM

This chapter is concerned with scheduling an assembly system with resource (assembly stations) constraints. Parts are completely machined before entering the assembly system. To illustrate the basic idea of the problem presented in this chater, consider a product shown in Figure 1(a). Since parts p1, p2, p3, and p4 have already been machined, the machining time of each part is 0. Then the problem can be viewed as consisting of subassemblies only, as shown in Figure 1(b).

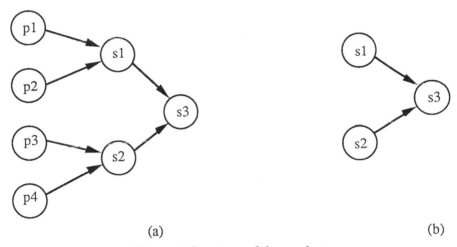

(a)   (b)

Figure 1. Structure of the product

Each subassembly requires a certain assembly station. For the two-product assembly problem in Figure 2, subassemblies s1, s2, s3, s4, s5, and s6 require stations t1, t3, t2, t1, t3, and t2 respectively. When many products are assembled in parallel, resource conflicts may occur. For example, concurrent execution of s1 and s4 would cause a resource conflict because both require station t1.

The designer of a product (assembly) has an impact on the structure of the assembly constraints and the number of assembly plans. These in turn influence schedulability of the assembly system. As an example, consider a product that involves four subassemblies. Three different structures of precedence constraints for the 4-subassemblies product are presented in Figure 3.

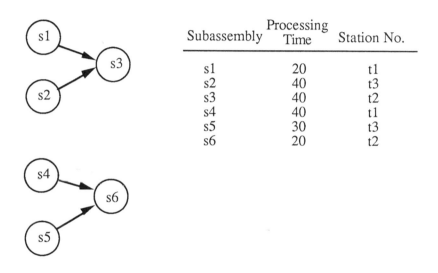

Subassembly	Processing Time	Station No.
s1	20	t1
s2	40	t3
s3	40	t2
s4	40	t1
s5	30	t3
s6	20	t2

Figure 2. A two-product assembly problem

For each of the three structures in Figure 3, the product could be assembled according to one assembly plan, called the basic assembly plan, or more than one assembly plan. Each assembly plan, in addition to the basic plan, is referred to as an alternative assembly plan.The alternative assembly process plan involves resources that are at least partially different from the resources in the basic assembly plan.

Then, for each structure, three different problems are considered:

1) the fixed assembly plan problem

2) the N-alternative assembly plan problem with processing times identical to the corresponding processing times in the basic assembly plan
3) the N-alternative assembly plan problem with processing times different than the corresponding processing times in the basic assembly plan

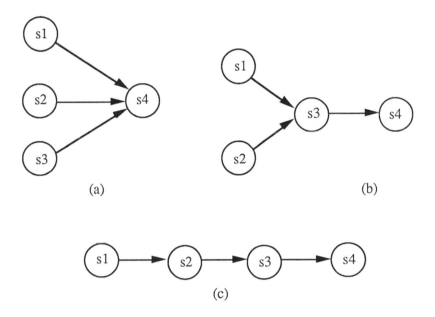

(a)

(b)

(c)

Figure 3. Three different structures of precedence constraints for a product with four subassemblies:
(a) parallel precedence constraints, (b) mixed (parallel and serial) precedence constraints, (c) serial precedence constraints

Each of the three above scheduling problems involves a large number of constraints. The two algorithms presented in this paper take advantage of the latter feature of the scheduling problem.

III. THE FIXED-ASSEMBLY PLAN PROBLEM

In the case when fixed-assembly plans are considered, the scheduling problem becomes a project scheduling problem and detailed scheduling is performed by an algorithm for general resource-constrained project scheduling. The latter approach seem to naturally fit the problem considered in this paper due to the large number of precedence constraints. Here a problem is represented as an "activity on nodes" (hereafter AON) network. Nodes represent subassemblies (activities), while the arcs represent required precedence relationships between subassemblies. The network is assumed to contain a unique dummy beginning and dummy finish subassembly, each with the processing time 0. Figure 4 shows also an AON network of the two-product problem in Figure 2. The information in Table 4 shows processing times and resource required by subassemblies. All subassemblies are assumed to have integer-valued processing times.

For a given precedence network, the unique schedule which assigns the earliest start time to each subassembly is computed. Since it is assumed that all subassemblies have integer-valued processing times, each subassembly's earliest start time is an integer number. Each time interval $(t, t+1)$ is examined to check for the existence of a resource violating set (RVS) of subassemblies where: (1) all subassemblies in RVS are executed concurrently under the current schedule, and (2) all these subassemblies require the same station. This algorithm plans from time 0 forward and looks for the smallest time t such that a resource violating set of subassemblies are in progress in $(t,t+1)$.

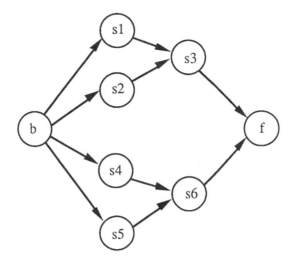

Subassembly No.	Station No.	Processing Time	Subassembly No.	Station No.	Processing Time
b*	--	0	s4	t1	40
s1	t1	20	s5	t3	30
s2	t3	40	s6	t2	20
s3	t2	40	f**	--	0

* b : beginning ** f : finish

Figure 4. AON network for the two-product problem

When a RVS is found, the algorithm attempts to resolve it by adding additional precedence constraints. For example, if all members of the RVS of subassemblies {s5 s6 s7} are in progress in the time interval (t,t+1) under the current schedule, six arcs {s5->s6, s5->s7, s6->s7, s6->s5, s7->s5, s7->s6} are to be considered. Each of these arcs expresses a constraint that one specific subassembly in {s5 s6 s7} must finish before a different specific subassembly in {s5 s6 s7} can start. Adding any one of these arcs to the precedence network blocks the concurrent execution of {s5 s6 s7}. In adding such arcs we use the "weakly-constrained approach", as in Bell and Park [1]. The weakly-constrained approach can be viewed as using the following "weak alternative to adding arc a->b". A dummy subassembly z is added with processing times $d_z = t_a + d_a$. Subassembly z requires no resources and can be assigned start time 0. Arcs are added from the dummy beginning subassembly to z and from z to b. The immediate impact of this change to the network is to increase the scheduled start time of b to $t_a + d_a$. The effect of this is then propagated across all arcs emanating from b and this propagation process continues in an obvious way until it dies out. However, there is no arc added to the network which connects any pair of subassemblies. Thus no such arc will participate in propagation in later decisions. The difference between the weakly-constrained approach and the approach presented in this paper is that when a subassembly is pushed forward by adding a weaker constraint, a station which can perform it in the shortest time is assigned. Thus a feasible schedule is generated by successively constraining the problem until no resource conflicts remain.

Based on the above discussion, Algorithm 1 is developed.

Algorithm 1 (The fixed-assembly plan problem)

Step 1.   Set up the precedence network.
               Assign the earliest start time to each subassembly in the
               network.

Step 2.   Look for the first resource violating set (RVS).
               If no RVS in the network, then report a feasible schedule and
               stop;
               otherwise, go to Step 3.

Step 3.   The resolution of RVS.

Disjunctive arcs each of which connecting one
subassembly in RVS to another subassembly in RVS.
Select an arc which causes the minimal immediate increase in
the makespan.
Add the selected arc to the network.
Propagate the effects through the network.
Go to Step 2.

Algorithm 1 is illustrated in Example 1.

Example 1

Find the minimum makespan schedule for a two-product problem in Figure
2.

Step 1.   Setup the precedence network (as shown in Figure 4).
                 The earliest start times are as follows:

Subassembly No.	s1	s2	s3	s4	s5	s6	f
start-time	0	0	40	0	0	40	80
finish-time	20	40	80	40	30	60	80
station	t1	t3	t2	t1	t3	t2	--

Step 2.   Find RVS {s2, s5} at 0, go to Step 3.
Step 3.   Consider the following alternatives:
                 (1) adding s2->s5: causes makespan to increase by 10
                 (2) adding s5->s2: causes makespan to increase by 30
                 Thus select and add to the network arc s2->s5 (alternative (1)
                 above).
                 Now the earliest start times are updated as follows:

Subassembly No.	s1	s2	s3	s4	s5	s6	f
start-time	0	0	40	0	40	70	90
finish-time	20	40	80	40	70	90	90
station	t1	t3	t2	t1	t3	t2	--

Go to Step 2.

Step 2.   Find RVS {s1, s4} at 0, go to Step 3.

Step 3.   Consider the following alternatives:

(1) adding s1->s4: causes makespan to increase by 0

(2) adding s4->s1: causes makespan to increase by 10

Thus select and add to the network arc s1->s4.

Now the earliest start times are updated as follows:

Subassembly No.	s1	s2	s3	s4	s5	s6	f
start-time	0	0	40	20	40	70	90
finish-time	20	40	80	60	70	90	90
station	t1	t3	t2	t1	t3	t2	--

Go to Step 2.

Step 2.   Find RVS {s3, s6} at 70, go to Step 3.

Step 3.   Consider the following alternatives:

(1) adding s3->s6: causes makespan to increase by 10

(2) adding s6->s3: causes makespan to increase by 10

Arbitrarily select and add to the network arc s3->s6.

Now the earliest start times are updated as follows:

Subassembly No.	s1	s2	s3	s4	s5	s6	f
start-time	0	0	40	20	40	80	100
finish-time	20	40	80	60	70	100	100
station	t1	t3	t2	t1	t3	t2	--

Go to Step 2.

Step 2. No RVS in the network, report the following schedule and stop.

Subassembly No.	s1	s2	s3	s4	s5	s6	f
start-time	0	0	40	20	40	80	100
finish-time	20	40	80	60	70	100	100
station	t1	t3	t2	t1	t3	t2	--

IV. THE N-ASSEMBLY PLAN PROBLEM

In the N-assembly plan problem, each subassembly can be performed on more than one station. The nature of resource versatility is exploited in resolving resource conflicts. Whenever it is possible, an attempt is made to shift a subassembly to the left as much as possible by switching the station assignment from the current station to a station which is idle. The idea is partly due to "left-shifting" in Schrage [5]. This "left-shifting" can not only resolve resource conflicts but also decrease the makespan.

The algorithm starts scheduling with the best station routing, i.e., a station which can perform that subassembly in the shortest time. For such a station routing, duration and resource-usage of each subassembly are determined and then an unique resource-constrained project scheduling problem can be identified.

Based on the above discussion, Algorithm 2 is developed.

Algorithm 2 (The N-assembly plan problem)

Step 1:   Set up the precedence network with the best station routing.
                 Assign the earliest start time to each subassembly in the
                 network.

Step 2:   Look for the first resource violating sets (RVSs).
                 If no RVS, then report a feasible schedule and stop.
                 Otherwise, go to Step 3.

Step 3:   The resolution of RVS (to decide which subassembly will be
                 assigned to
                 alternative station or will be delayed by adding a disjunctive
                 arc).
                 For each subassembly A in RVS, consider:
                         (1) alternative station assignments of subassembly A
                         (2) disjunctive arcs each of which connecting a
                             subassembly in RVS to subassembly A.
                 Select a case which causes the minimal increase in the
                 makespan.
                 Update the network by the selected choice:
                 Propagate the effects through the network.
                 Go to Step 2.

Algorithm 2 is illustrated in Example 2.

Example 2

Find the minimum makespan schedule for the problem in Figure 5. Now
each subassembly can be performed on more than 1 station. For example,
subassembly s1 can be done either on station t1 or on station t2.

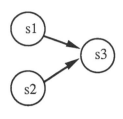

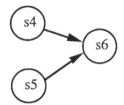

Subassembly	Station No.		
	t1	t2	t3
s1	20	22	--
s2	44	--	40
s3	--	40	44
s4	40	--	44
s5	--	33	30
s6	22	20	--

Figure 5. Two products with alternative assembly plans

Step 1.   Setup the precedence network with the best station routing (same as shown in Figure 4)

The earliest start times are as follows:

Subassembly No.	s1	s2	s3	s4	s5	s6	f
start-time	0	0	40	0	0	40	80
finish-time	20	40	80	40	30	60	80
station	t1	t3	t2	t1	t3	t2	--

Step 2.   Find RVS {s2, s5} at 0, go to Step 3.

Step 3.   Consider the following alternatives:

(1) adding s2->s5: causes makespan to increase by 10

(2) switching t3 to t2 for s2: causes makespan to increase by 4

(3) adding s5->s2: causes makespan to increase by 30

(4) switching t3 to t2 for s5: causes makespan to increase by
    0

Thus select case (4) above .

Now the earliest start times are updated as follows:

Subassembly No.	s1	s2	s3	s4	s5	s6	f
start-time	0	0	40	0	0	40	80
finish-time	20	40	80	40	33	60	80
station	t1	t3	t2	t1	t2	t2	--

Go to Step 2.

Step 2.   Find RVS {s1, s4} at 0, go to Step 3.

Step 3.   Consider the following alternatives:

(1) adding s1->s4: increases the makespan by 0

(2) switching t1 to t2 for s1: impossible

(3) adding s4->s1: increases the makespan by 10

(4) switching t1 to t3 for s4: impossible

Thus select case (1).

Now the earliest start times are updated as follows:

Subassembly No.	s1	s2	s3	s4	s5	s6	f
start-time	0	0	40	20	0	40	80
finish-time	20	40	80	60	33	60	80
station	t1	t3	t2	t1	t2	t2	--

Go to Step 2.

Step 2.   Find RVS {s3, s6} at 60, go to Step 3.

Step 3.   Consider the following alternatives:

(1) adding s3->s6: increases the makespan by 20

(2) switching t2 to t2 for s3: increases the makespan by 4

(3) adding s6->s3: increases the makespan by 10

(4) switching t2 to t1 for s6: increases the makespan by 2

Select case (4).

Now the earliest start times are updated as follows:

Subassembly No.	s1	s2	s3	s4	s5	s6	f
start-time	0	0	40	20	0	60	82
finish-time	20	40	80	60	33	82	82
station	t1	t3	t2	t1	t2	t1	--

Go to Step 2.

Step 2.   No RVS in the network, report the following schedule and stop.

Subassembly No.	s1	s2	s3	s4	s5	s6	f
start-time	0	0	40	20	0	60	82
finish-time	20	40	80	60	33	82	82
station	t1	t3	t2	t1	t2	t1	--

V. COMPUTATIONAL EXPERIENCE

For each of three different structures of precedence constraints defined in the Section 2, four different 5-part problem types with 4 stations are constructed:

(1+0) : the fixed assembly plan problem

(1+1,0): the 1-alternative assembly plan problem with processing times identical to the corresponding processing times in the basic assembly plan

(1+1,5): the 1-alternative assembly plan problem with processing times 5% higher than the corresponding processing times in the basic assembly plan

(1+1,10): the 1-alternative assembly plan problem with processing times
         10% higher than the corresponding processing times in the basic
         assembly plan

As shown in Figure 6, 12 different problem types are dealt with in this
paper. In order to generalize the computational results, a set of random test
problems was constructed. In this set, the processing time and the station
required of each subassembly were randomly assigned. The processing time
varied between 20 and 40. The products were randomly assigned to stations
1 through 4 . The performance measures were averaged for each problem
type (see Tables 1 through 3).

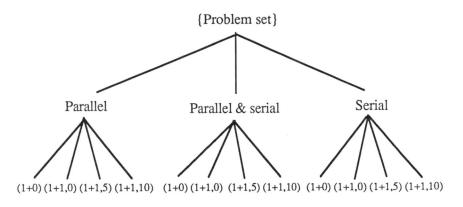

Figure 6. Structure of the test problems

Computations were performed on an Apple Macintosh SE using a version of
Common Lisp (Allegro Common Lisp). Several performance measures such
as makespan, station utilization ratio, product mean flow time, and the CPU
time were computed. The summarized computational results are shown in
Tables 1 through 3.

Table 1. Parallel precedence constraints

Type	Makespan	Station utilization				Product mean flow time	CPU time (sec.)
		t1	t2	t3	t4		
(1+0)	238.4	.73	.52	.57	.74	193.1	34
(1+1,0)	192.5	.85	.74	.73	.81	160.2	58
(1+1,5)	192.8	.86	.75	.73	.81	160.6	58
(1+1,10)	197.2	.83	.74	.73	.80	163.4	59

Table 2. Mixed (parallel and serial) precedence constraints.

Type	Makespan	Station utilization				Product mean flow time	CPU time (sec.)
		t1	t2	t3	t4		
(1+0)	239.7	.73	.51	.57	.73	195.9	26
(1+1,0)	190.4	.83	.76	.74	.84	161.4	32
(1+1,5)	193.5	.86	.72	.73	.83	163.4	33
(1+1,10)	194.5	.87	.71	.74	.84	165.8	33

Table 3. Serial precedence constraints

Type	Makespan	Station utilization				Product mean flow time	CPU time (sec.)
		t1	t2	t3	t4		
(1+0)	245.0	.71	.49	.56	.71	202.8	21
(1+1,0)	196.8	.81	.73	.75	.77	169.5	23
(1+1,5)	200.6	.84	.71	.72	.77	170.5	23
(1+1,10)	201.0	.81	.73	.74	.78	173.1	23

It is shown that the type of precedence constraints has an impact on the performance measures. In general, the assemblies with parallel precedence constraints are the most favorable; mixed (parallel and serial) precedence constraints are the next; serial precedence constraints have the worst performance, in all measures except the CPU time. However, in the CPU time, the opposite performance was observed.

The impact of the alternative assembly plans is stronger than the type of precedence constraints. The alternative assembly plan problem has the following advantages over the fixed assembly plan:

(1) Makespan is reduced.
(2) Station utilization becomes balanced among stations and enhanced.
(3) Product mean flow time is reduced.

For discussion of many other aspects of concurrent engineering see [8], [9], and [10].

VI. CONCLUSIONS

Decisions made in the assembly design phase can be judged based on many constraints. Schedulability is one of those constraints. There are many objectives regarding schedulability such as minimizing makespan, meeting due dates, minimizing the part flow time, and balancing assembly line. On the other hand, a scheduling time is critical. Although the scheduling time mainly depends on the algorithm used, it is also affected by certain decisions made in the design phase. Thus aiding the designer of product to make reasonable decisions from the scheduling standpoint is an important task.

This paper aims at addressing the importance of the assembly design on schedulability of assemblies. This is the first attempt to analyze the effects of the structure of precedence constraints and the number of assembly plans together. It was shown that the serial structure of precedence constraints always results in the inferior value of performance measures such as makespan, product mean completion time, and station utilization rate. However, the serial structure of precedence constraints requires the least computation time of the other two structures. The assemblies with the parallel precedences perform superior in terms of performance measures such as makespan, product mean completion time, and station utilization rate.

References

1. Bell, C.E. and Park, K., "Solving resource-constrained project scheduling problems by A* search", *Naval Research Logistics*, **37**, 64-84, 1990.
2. Donath, M. and Graves, R.J., "Flexible assembly systems: test results for an approach for near real-time scheduling and routing of multiple products", *International Journal of Production Research*, **27**, 215-227, 1989.
3. Fry, T.D., Oliff, M.D., Minor, E.D., and Leong, G.K., "The effects of product structure and sequencing rule on assembly shop performance", *International Journal of Production Research*, **27**, 671-686, 1989.
4. Ghosh, S. and Gagnon, R.J., "A comprehensive literature review and analysis of the design, balancing and scheduling of assembly systems", *International Journal of Production Research*, **27**, 637-670, 1989.
5. Schrage, L., "Solving resource-constrained network problems by implicit enumeration - non preemptive case", *Operations Research*, **18**, 263-278, 1970.
6. Wilhelm, W.E. and Shin, H., "Effectiveness of alternative operations in a flexible manufacturing system", *International Journal of Production Research*, **23**, 65-79, 1985.

7. Udomkesmalee N. and Daganzo, C., "Impact of parallel processing on job sequences in flexible assembly systems", *International Journal of Production Research*, **27**, 73-89, 1989.

8. Kusiak, "Intelligent Manufacturing Systems," Prentice Hall, Englewood Cliffs, N.J., 1990.

9. Kusiak, A. and K. Park, Concurrent Engineering: decomposition and scheduling of design activities, *International Journal of Production Research*, **28**, 1883-1900, 1990.

10. Kusiak, A. (Ed.), "Intelligent Design and Manufacturing," John Wiley, New York, 1992.

Multi-Criteria Optimization and Dynamic Control Methods in Flexible Manufacturing Systems

In-Kyu Ro

Joong-In Kim

CIMS Laboratory, Department of

Industrial Engineering, Hanyang University,

SEOUL 133-791, KOREA

I. INTRODUCTION

There are many methods to the scheduling and operational control problems in Flexible Manufacturing Systems (FMSs). These methods are mathematical programming techniques (optimization and heuristic methods), queueing networks, computer simulation, petri nets, and Expert Systems, etc. Most FMS models are based on one out of the several methods listed above and they are single-objective oriented techniques. However, the combination of the several methods may give better results than the single one and decision makers usually consider multiple objectives. On the other hand, most FMS models consider the scheduling for jobs and AGVs,[1]

---

[1]AGVs, Automated Guided Vehicles.

CONTROL AND DYNAMIC SYSTEMS, VOL. 47
Copyright © 1991 by Academic Press, Inc.
All rights of reproduction in any form reserved.

seperately. Jobs and AGVs are closely related and should be operated dependently. Thus the simultaneous scheduling for jobs and AGVs is strongly needed.

In this paper the scheduling and operational control problems for jobs and AGVs in an FMS having local buffers with limited capacities are considered. Such problems are divided into the following six sub-problems.

1) part type selection (input sequence) during initial entry

2) part type selection (input sequence) during general entry

3) process selection (routing mix or machine center selection)

4) part-to-machine allocation (dispatching or sequencing)

5) AGV dispatching

6) AGV route selection

Especially, in the process selection (routing mix or machine center selection) problem, the new optimization method using the combination of the Multiple-Objective Linear Programming (MOLP) and closed queueing network and the dynamic control method are developed. The new MOLP models consider workload balancing and some performance measures such as makespan, mean flow time, and due date. The closed queueing network (using CAN_Q) determines whether the workload balancing should be considered to the FMS or not and the performance measures considered are production rates, mean flow time, and system utilization for all machines. And then, the results of the static optimization method are revised by the dynamic and real time control method during the production. These methods can flexibly cope with the change of system configuration (machine breakdown, etc). Finally, these new methods are compared with the existing rules and methods by computer simulations. The simulation experiments include the six sub-problems, limited local buffer capacity, limited number of AGVs and pallets, input/output and flow controls, and several performance measures (such as makespan, mean flow time, mean tardiness, maximum tardiness, and system utilization) which are obtained simultaneously from each simulation.

This paper is organized as follows. In Section II, the six scheduling and operational control problems for jobs and AGVs are discussed and the new methods in the process selection problem are developed. In Section III, an example in an FMS having local buffers with limited capacities is presented. Simulation results and validation of the model are described in Section IV. Conclusions are presented in Section V.

II. THE SCHEDULING AND OPERATIONAL CONTROL PROBLEMS FOR JOBS AND AGVS

Assumptions involved in this paper are as follows :

(1) The sequence of operations (and hence a serial precedence relationship among operations) for each part type is known.

(2) Processing time for each operation is known.

(3)   AGV travel distance from machine to machine is known.

(4)   Local buffer capacity is limited and the number of AGVs
      are fixed.

(5)   Demand for each part type arrives at different times.

(6)   Raw materials are available at time zero.

The six scheduling and operational control sub-problems for jobs (part types) and AGVs are discussed below.

A. Part Type Selection (Input Sequence) During Initial Entry

The parts initially enter an empty and idle system during the start-up stage. If the system reaches a quick steady state or continues in it for a long time, the effect of the initial entry order of different part types on the performance of the system may not be significant. However, the effect of initial entry order will be significant for a system which is emptied periodically, e.g. for preventive maintenance, or the system empties at the end of every shift [1]. In an FMS, an initial entry period will cover the time period until the system become a loaded system. A loaded system is defined as a system which can hold up to N parts on pallets at a machine, or recirculating after the first N parts have been entered [2].

In this paper, EDD (Earliest Due Date) rule is used for the simulation experiments as a part type selection rule during initial entry.

B. Part Type Selection (Input Sequence) During General Entry

After   the   initial   entry stage is over and as   the   parts   go through   the processing,   the next step would be the   ordering   of subsequent parts to enter the system.

In an FMS having local buffers with limited capacities,   parts are released when a machine which can process the first operation for each part type is idle or when the local buffer of that machine has remaining spaces. In this paper, the following Highest Ratio of Remaining Requirement rule is selected randomly for the simulation experiments :

The part having highest ratio of remaining requirement to their original requirement will be selected first.

C. Process Selection (Routing Mix or Machine Center Selection)

A part after entering will visit machines described by its fixed (pre-determined) process routing or dynamically determined by some decision criteria. In this paper, existing static and

Table I. Summary of process selection rules and methods

	Existing method	New method
Static method	NAR ARP	MOLP Optimization - MARP
Dyamic method	ARD ARPD WINQ	Dynamic Control Method - MARPD

dynamic rules and methods [3] are described and then the new
multi-criteria optimization and dynamic control methods are
suggested. Table I shows the summary of the rules and methods
described in this section.

1. The Existing Rules and Methods

a. NAR (No Alternative Routings or Minimum Total Processing Time Rule)

If no alternative routings are permitted, the route with the
minimum total processing time is selected and each part must be
processed according to a fixed sequence of operations. However,
marked imbalance may occur among the workloads assigned to each
machine and the system can not react to machine breakdowns in this
rule.

b. ARP (Alternative Routings Planned)

Alternative routings may be planned using a LP model which
holds the objective of minimizing makespan. The LP model can be
formulated as follows :

Min. M

s.t. Production requirement for each part type

$$\sum_j X_{ij} \geqslant N_i \qquad\qquad \forall i \qquad\qquad (1)$$

Workload assigned to individual machines

$$\sum_i \sum_j t_{ijk} X_{ij} \leqslant M \qquad\qquad \forall k \qquad\qquad (2)$$

Due date for each part type

$$\sum_j \sum_k (t_{ijk} + T_{ij}) X_{ij} + A_i \leqslant D_i \qquad \forall i \qquad\qquad (3)$$

$$X_{ij} \geqslant 0, \ M \geqslant 0$$

where

  i : part type (i = 1,2,...,I)

  j : alternative route (j = 1,2,...,J)

  k : machine (k = 1,2,...,K)

  M : makespan

X$_{ij}$: number of part type i to be produced in alternative route j

N$_i$: total requirement of part type i

t$_{ijk}$: processing time of part type i on machine k in alternative route j

T$_{ij}$: travel time of part type i in alternative route j

A$_i$: arrival time of part type i

D$_i$: due date of part type i

In the solution (routing mix) of the LP model above, part types can follow different routings. However, switching between alternative routings may not be allowed once the part is released. In addition, the solution is absolute minimum since queueing time and precedence relationships are not taken into account in the model. Furthermore, if one wants to model tooling requirements, one may be confronted with nonlinearlities.

The problem above can be formulated as Integer Programming but

it may require prohibitive computational time so that (rounded)
LP solutions are used to reduce the run time.

A series of LPs needs to be solved over time as the system
configuration changes due to schedule adjustments, to new (or
sudden) job arrivals, or in response to machine breakdowns.

c. ARD (Alternative Routings Directed Dynamically)

This dynamic rule is to deliver the current part to the next
machine which has the shortest [travel time + queueing time +
operation time] for that part among the candidate machines. In
this rule, queueing time is defined as the remaining operation
time for the part currently being processed by the candidate
machine plus operation times for the waiting parts with higher
priority at the local buffer of the candidate machine than the
priority of this part.

If a machine breakdown occurs, take that failed machine out
from the available machine list during repair time. The current
part being processed by that machine is transferred to alternative
machine according to the same criterion above. In this case, new
operation time at the alternative machine is defined as follows :
New operation time

= (Remaining operation time at the failed machine x
Original operation time at the alternative machine) /
Original operation time at the failed machine

d. ARPD (Alternative Routings Planned and Directed Dynamically)

This dynamic method extends the previous ARP method to allow certain routing decisions to be made dynamically. Goals for production of alternative routes, X_{ij}, are still observed but the control is permitted to the status of the system as it evolves over time.

The solution procedures of ARPD method are as follows :

Step 1.

Solve the LP problem.

If primary machine (LP solution) is idle, select that machine.

Otherwise, go to step 2.

Step 2.

If primary machine is busy and only one alternative machine is idle, select that alternative machine.

If primary machine is busy and two or more alternative machines are idle, select the alternative machine which has the shortest [travel time + operation time] for that part.

Otherwise, go to step 3.

Step 3.

Select the machine which has the shortest [travel time + queueing time + operation time] for that part.

As stated above, a series of LPs needs to be solved over time as the system configuration changes due to schedule adjustments, to new (or sudden) job arrivals, or in responce to machine breakdowns.

e. WINQ (Work IN Queue)

A machine which has the least work in queue in terms of processing time is selected.

2. The New Methods

Fig. 1 shows the solution procedures for the new methods. The solution procedures are explained below.

a. The New MOLP (Multiple Objective Linear Programming) Model

Alternative routings may be planned using a single-objective LP model and different LP models are appeared in the several previous literatures [3,4,5,6,7]. In this paper, a new MOLP model that considers several performance measures and workload balancing is presented. The new MOLP model can be formulated as follows :

Min. M (Makespan)

Min. $\sum_i (C_i - A_i) / I_i$ (Mean Flow Time)

s.t.

Production requirement for each part type

$$\sum_j X_{ij} \geqslant N_i \qquad\qquad \forall i \qquad\qquad (4)$$

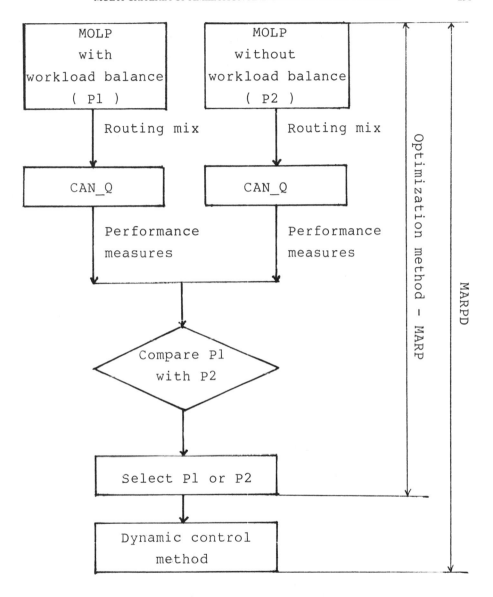

Fig.1. The diagram of the solution procedures for the new methods

* Legends :
  MOLP = Multiple Objective Linear Programming
  MARP = Modified Alternative Routings Planned
  MARPD = Modified Alternative Routings Planned and Directed
          Dynamically

Workload assigned to individual machines

$$\sum_i \sum_j t_{ijk} X_{ij} \leqslant M \qquad \forall k \qquad\qquad (5)$$

Completion time for each part type

$$\sum_j \sum_k (t_{ijk} + T_{ij}) X_{ij} + A_i \leqslant C_i \qquad \forall i \qquad (6)$$

Due date for each part type

$$C_i \leqslant D_i \qquad\qquad \forall i \qquad\qquad (7)$$

Makespan

$$C_i \leqslant M \qquad\qquad \forall i \qquad\qquad (8)$$

Workload balance for each machine

$$(\sum_i \sum_j \sum_k t_{ijk} X_{ij}) / K - \sum_i \sum_j t_{ij} X_{ij} \geqslant 0 \qquad \forall k \qquad (9)$$

$$X_{ij} \geqslant 0, \ M \geqslant 0, \ C_i \geqslant 0$$

where the symbols used represent the following :

    i : part type (i = 1, 2, ..., I)

    j : alternative route (j = 1, 2, ..., J)

    k : machine (k = 1, 2, ..., K)

    M : makespan

   X_{ij} : number of part type i to be produced in alternative route j

   N_i : production requirement of part type i

   t_{ijk} : processing time of part type i on machine k in alternative route j

   T_{ij} : travel time of part type i in alternative route j

   A_i : arrival time of part type i

C_i : completion time of part type i

D_i : due date of part type i

The new MOLP model above considers the following performance measures.

1) minimize makespan (implicitly , maximize production rate and maximize average machine utilization)

2) minimize mean flow time (implicitly, minimize average WIP level, minimize mean lateness and minimize mean waiting time)

3) measures related to due date (mean tardiness, maximum tardiness and number of late jobs)

4) workload balancing (i.e. machine utilization balancing)

" Minimize makespan " is equivalent to " maximize production rate " and " maximize average machine utilization " [8]. Also, the WIP level could be affected by the makespan [9].

" Minimize mean flow time " is equivalent to " minimize average WIP level ", " minimize mean tardiness ", and " minimize mean waiting time " [10].

Constraints (7) assume that due date for each part type must be met and hence the value of performance measures related to due date are zero's. Constraints (9) balance the workload for each machine and are concerned with preventing a machine from being a bottleneck. Even though workload balancing is not necessarily the best objective in an FMS, it can be an important objective in some practical situations because of discreteness of operation times, different machine tool requirements, and limited capacity tool magazines [11]. However, the workload balancing might conflict with the minimization of makespan [9]. Thus, we feel the trade-off between workload balancing and the other performance measures should be investigated. In addition, we must determine whether the workload balancing will be introduced to the FMS or not. This type of decision making is solved by the optimization method suggested in the following section.

b. Optimization Method - MARP (Modified Alternative Routings Planned)

Ro and Kim [3] developed three new process selection rules and methods (ARD, ARP, and ARPD) in process selection problem for an FMS that has local buffers with limited capacities. Among those rules and methods, ARP uses an LP model which holds the objective of minimizing makespan and gives an absolute optimal solution. In this paper an optimization approach is suggested. The approach modifies the ARP and consists of a hierachical combination of a newly developed MOLP model and a queueing network model. The limitation of the MOLP model is that queueing time is not taken into account in the model and informations on the dynamic working of an FMS can not be provided. Thus, the solution of MOLP model is then examined by a queueing network model considering some dynamic effects (such as waiting lines and transport systems).

The solution procedures of optimization approach are as follows:

Step 1.

Solve the MOLP model with workload balancing constraints (P1) and the MOLP model without workload balancing constraints (P2).

From the two MOLP solutions, the routing mixs for all part types are obtained, respectively.

Step 2.

Given the routing mixs from the two MOLP solutions, use the CAN_Q to obtain some performance measures, such as production rates, mean flow time, and machine utilizations, etc.

Step 3.

Compare P1 and P2 based on the values of performance measures obtained from CAN_Q.

Choose the best one between the solution (routing mix) of P1 and that of P2. This implies decision making of determination whether the workload balancing is applied to the FMS or not.

There are a large number of solution methods for the multiple-objective linear model. The three primary methods or philosophies are [12] :

   i) Weighting or utility methods

  ii) Ranking or prioritizing methods

iii) Efficient solution (or generating) methods

In addition, another formulation and solution method different from the MOLP model suggested in this paper can be possible by the Goal Programming model.

CAN_Q is fully described in some references [13,14,15].

c. Dynamic Control Method - MARPD (Modified Alternative Routings Planned and Directed Dynamically)

This dynamic control method revises the existing ARPD method. That is, the solution by the LP model in the ARPD method is replaced by the solution obtained from the optimization method and the solution procedures of the dynamic control method are same as the ARPD.

D. Part-To-Machine Allocation (Dispatching or Sequencing)

After selecting the next machine which will process next operation, a part visits the machine by an AGV. If the machine is idle, its processing will start without delay. However, if the machine is working on some other part, the part will join the queue. When the machine finishes the current work, a new part will be selected from the queue of waiting parts.

In this paper, SPT (Shortest Processing Time) rule is selected randomly for the simulation experiments.

E. AGV Dispatching

When a part is completed at the current machine, that part must wait at the local buffer of that machine until both an AGV and the local buffer space of a next selected machine become available.

The steps of the AGV dispatching rule are as follows :

Step 1.

If there is only one idle vehicle, select that vehicle.

If two or more idle vehicles are available, select the nearest vehicle among them.

Otherwise, wait until an vehicle is available and go to step 2.

Step 2.

If only one pick-up request exists, select that request.

If two or more pick-up request exist, select the longest waiting pick-up request except the case that there exists the top priority part which has completed the last operation and hence can exit the system immediately.

Otherwise, AGV stays at the place where it was made idle and waits for pick-up request.

F. AGV Route Selection

1. Uni-directional Guide Path

A uni-directional flow model is to determine the shortest path between all the points in the layout and AGVs travel along that route at all times. The same routes between all the points are taken regardless of the congestion status of the routes. In this case, traffic control is rather simple but travel time is longer than the bi-directional flow.

2. Bi-directional Guide Path

A bi-directional flow model is more flexible and travel time is shorter than the uni-directional flow model, but the AGV conflict prevention is important.

The AGV route selection rule which prevents the AGV conflict is as follows : An AGV will attempt to take the shortest path to its destinations. However, in situations where one or more vehicles are involved in head-on conflict (i.e., at some control point, if next segment along the shortest path is occupied by the other vehicle travelling to this control point), alternative non-occupied segment is selected dynamically to avoid the conflict.

III. EXAMPLE IN AN FMS

A. System Description and Input Data

The hypothetical FMS is assumed to be composed of four machining centers (MCs), finite buffer space at each MC, a load/unload (L/U) station, and two AGVs which can carry one pallet at a time. AGV guide path is bi-directional. The number of pallets in the system is 28 and AGV speed is 45m/min. In order to investigate the effect of buffer capacity, the model is tested with two buffer capacities, 3 and 5 (q = 3, q = 5) and with two AGVs, 1 and 2.

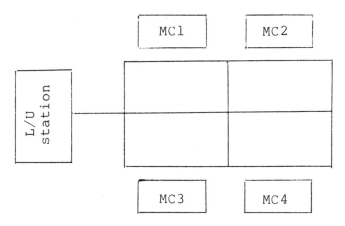

Fig.2. Schematic diagram of the hypothetical FMS

Since local buffer capacities are limited, there is a possibility that machine blocking occurs. In this paper, input/output and flow control are used to avoid the blocking of machine. For the input control, a part may be entered to the system if and only if the total number of parts in the FMS is less than the number of pallets available in the FMS [16]. For the flow control, when the local buffer of the selected machine which can process the next operation for each part type has more than 1 part for the case of q = 3 and 3 parts for the case of q = 5, any part would not capture an AGV. These control limits are determined by several simulation tests. For the output control, the part which has completed the last operation has the top pick-up priority and hence exit the system immediately [16].

Table II gives the arrival time, the alternative routes, the sequence of operations along with unit processing time, the shortest route travel times, and the total processing times for each part type.

Three different part types, A, B, and C are defined as shown in Table II. Each requires four operations and has to visit each of

Table II. Input data for each part type

Part type	Arrival time	Alter- native routes	Sequence position				Route travel	Total process time
			1	2	3	4		
A	10	1	1(4)	3(7)	1(3)	4(2)	2.0	18.0
		2	1(4)	3(7)	2(5)	4(2)	2.0	20.0
		3	2(6)	3(7)	1(3)	4(2)	2.2	20.2
		4	1(4)	3(7)	3(7)	4(2)	1.3	21.3
		5	2(6)	3(7)	2(5)	4(2)	2.2	22.2
		6	2(6)	3(7)	3(7)	4(2)	1.8	23.8
B	20	1	2(4)	1(3)	2(1)	4(10)	1.8	19.8
		2	2(4)	1(3)	2(1)	1(12)	1.6	21.6
		3	2(4)	1(3)	3(3)	4(10)	1.8	21.8
		4	3(7)	1(3)	2(1)	4(10)	2.0	23.0
		5	2(4)	1(3)	3(3)	1(12)	2.0	24.0
		6	3(7)	1(3)	2(1)	1(12)	1.8	24.0
		7	3(7)	1(3)	3(3)	4(10)	2.0	25.0
		8	3(7)	1(3)	3(3)	1(12)	2.2	27.2
C	30	1	1(5)	3(1)	3(1)	2(1)	1.6	9.6
		2	1(5)	4(2)	3(1)	2(1)	1.8	10.8
		3	1(5)	3(1)	4(3)	2(1)	1.8	11.8
		4	1(5)	4(2)	4(3)	2(1)	1.6	12.6
		5	1(5)	2(4)	3(1)	2(1)	2.0	13.0
		6	1(5)	2(4)	4(3)	2(1)	2.0	15.0

* The figures in parentheses are the unit processing times.

the MCs, but indifferent sequences. The set of alternative operations allows workloads to be transferred among all MCs to achieve the balance sought by the ARP, ARPD, MARP and MARPD in the process selection problem. It is assumed that it takes 3 minutes to load and unload a part on a pallet at L/U station. The time to put the pallet on or to take it off the AGV is assumed to be 0.8 minutes.

Table III gives the production requirement, the longest total processing time (which is the production requirement times the longest route processing time among the several alternative routes), and the due date for each production cycle and each part type.

When each part type arrives, due date is assigned based on the TWK (Total WorK content) rule. In the previous papers [17,18], this rule has been found to be quite effective for setting due date. According to this rule, due date D_i assigned to part type i is given by,

$$D_i = A_i + F\ P_i$$

where

A_i : arrival time for part type i

F : flow allowance ($F \geqslant 1$)

P_i : total processing time for part type i

In a conventional job shop, since there was a fixed route without alternative routings for each part type, total processing time is easily estimated. However, in an FMS, several alternative

Table III. Production requirement and due date for each part type.

Production cycle	Part type	Production requirement	Longest total processing time	Due date
1	A	8	190.4	200
	B	6	163.2	183
	C	12	180.0	210
2	A	16	380.8	391
	B	12	326.4	346
	C	24	360.0	390
3	A	32	761.6	772
	B	24	652.8	673
	C	48	720.0	750
4	A	60	1428.0	1438
	B	45	1224.0	1244
	C	90	1350.0	1380
5	A	80	1904.0	1914
	B	60	1632.0	1652
	C	120	1800.0	1830
6	A	100	2380.0	2390
	B	75	2040.0	2060
	C	190	2250.0	2280

routings exist and hence several total processing times for each part type and criterion for selecting total processing time has not existed until now.

Since queueing time is not embedded in the input data in Table II , the longest total processing time with the flow allowance $F = 1$ is selected.

B. ARP Method to the FMS

In an attempt to identify the number of the alternative routes for each part type, the LP model of the problem for the first production cycle can be formulated as follows :

Min. M

s.t.

Production requirement for each part type

$$X_{11} + X_{12} + X_{13} + X_{14} + X_{15} + X_{16} \geqslant 8$$

$$X_{21} + X_{22} + X_{23} + X_{24} + X_{25} + X_{26} + X_{27} + X_{28} \geqslant 6$$

$$X_{31} + X_{32} + X_{33} + X_{34} + X_{35} + X_{36} \geqslant 12$$

Workload assigned to individual machines

$$7X_{11} + 4X_{12} + 4X_{13} + 3X_{14} + 3X_{21} + 15X_{22} + 3X_{23} + 15X_{24}$$
$$+ 3X_{25} + 15X_{26} + 3X_{27} + 15X_{28} + 5X_{31} + 5X_{32} + 5X_{33} + 5X_{34}$$
$$+ 5X_{35} + 5X_{36} \leqslant M$$

$$5X_{12} + 6X_{14} + 11X_{15} + 6X_{16} + 5X_{21} + 5X_{22} + 4X_{23} + 4X_{24}$$
$$+ X_{25} + X_{26} + X_{31} + X_{32} + X_{33} + X_{34} + 5X_{35} + 5X_{36} \leqslant M$$

$$7X_{11} + 7X_{12} + 14X_{13} + 7X_{14} + 7X_{15} + 14X_{16} + 3X_{23} + X_{24}$$
$$+ 7X_{25} + 7X_{26} + 10X_{27} + 8X_{28} + 2X_{31} + X_{32} + X_{33} + X_{34} \leqslant M$$

$$2X_{11} + 2X_{12} + 2X_{13} + 2X_{14} + 2X_{15} + 2X_{16} + 10X_{21} + 12X_{22}$$
$$+ 10X_{24} + 10X_{27} + 3X_{32} + 2X_{33} + 5X_{34} + 3X_{36} \leqslant M$$

Due date for each part type

$$18.0X_{11} + 20.0X_{12} + 21.3X_{13} + 20.2X_{14} + 22.2X_{15} + 23.8X_{16}$$
$$+ 10 \leqslant 200$$

$$19.8X_{21} + 21.6X_{22} + 23.8X_{23} + 22.0X_{24} + 23.0X_{25} + 24.8X_{26}$$

$$+ 25.0X_{27} + 25.2X_{28} + 20 \leqslant 183$$

$$9.6X_{31} + 11.8X_{32} + 10.8X_{33} + 12.6X_{34} + 13.0X_{35} + 15.0X_{36}$$

$$+ 30 \leqslant 210$$

$$X_{ij} \geqslant 0 \quad \forall\, i, j \quad , \quad M \geqslant 0$$

Table IV shows the results of the LP model which describe the number of parts routed according to alternative routes for each part type.

Table IV. Results of the LP model for six production cycles.

Produc-tion cycle	Part type	1	2	3	4	5	6	7	8	Sum	Total
		\multicolumn{8}{c}{Number of parts routed according to alternative routes}		
1	A	8	0	0	0	0	0	0	0	8	
	B	3	0	1	0	2	0	0	0	6	26
	C	0	0	0	1	0	11	0	0	12	
2	A	16	0	0	0	0	0	0	0	16	
	B	5	0	2	0	5	0	0	0	12	52
	C	0	0	0	1	0	23	0	0	24	
3	A	32	0	0	0	0	0	0	0	32	
	B	11	0	3	0	10	0	0	0	24	104
	C	0	0	0	2	0	46	0	0	48	
4	A	60	0	0	0	0	0	0	0	60	
	B	20	0	6	0	19	0	0	0	45	195
	C	0	0	0	4	0	86	0	0	90	
5	A	80	0	0	0	0	0	0	0	80	
	B	27	0	8	0	25	0	0	0	60	260
	C	0	0	0	6	0	114	0	0	120	
6	A	100	0	0	0	0	0	0	0	100	
	B	33	0	10	0	32	0	0	0	75	325
	C	0	0	0	7	0	143	0	0	150	

C. MARP Method to the FMS

In this example, we can assume that decision maker gives the same weights for the two objectives (min. makespan and min. mean flow time). Thus, the weighting method with equal weights (w_1 = w_2 = 0.5) is used to solve the MOLP model because it is attractive from a strictly computational point of view (e.g. conventional simplex method may be used if the model is linear). That is, the multiple-objective model is transformed into a single-objective model and solved by the simplex method. The new objective function is :

$$\text{Min. } w_1 M + w_2 \sum_i (C_i - A_i) / I_i$$
$$= \text{Min. } 0.5 M + 0.5 \sum_i (C_i - A_i) / I_i$$

MOLP model with workload balancing (P1) :

```
MIN. M+0.3333333C1+0.3333333C2+0.3333333C3-10
S.T.
X11+X12+X13+X14+X15+X16>=8
X21+X22+X23+X24+X25+X26+X27+X28>=6
X31+X32+X33+X34+X35+X36>=12
7X11+4X12+3X13+4X14+3X21+15X22+3X23+3X24+15X25+15X26+3X27+15X28
+5X31+5X32+5X33+5X34+5X35+5X36-M<=0
5X12+6X13+11X15+6X16+5X21+5X22+4X23+1X24+4X25+X26+X31+X32+X33
+X34+5X35+5X36-M<=0
7X11+7X12+7X13+14X14+7X15+14X16+3X23+7X24+3X25+7X26+10X27+10X28
+2X31+X32+X33+X35-M<=0
2X11+2X12+2X13+2X14+2X15+2X16+10X21+10X23+10X24+10X27+2X32+3X33
+5X34+3X36-M<=0
18.0X11+20.2X12+20.2X13+21.3X14+22.2X15+23.8X16-C1<=-10
19.8X21+21.6X22+21.8X23+23.0X24+24.0X25+24.8X26+25.0X27+27.2X28
-C2<=-20
9.6X31+10.8X32+11.8X33+12.6X34+13.0X35+15.0X36-C3<=-30
C1<=200
C2<=183
C3<=210
```

```
C1-M<=0
C2-M<=0
C3-M<=0
16X11+18X12+18X13+20X14+20X15+22X16+18X21+20X22+20X23+21X24
+22X25+23X26+23X27+25X28+8X31+9X32+10X33+11X34+11X35+13X36
-28X11-16X12-12X13-16X14-12X21-60X22-12X23-12X24-60X25-60X26
-12X27-60X28-20X31-20X32-20X33-20X34-20X35-20X36>=0
16X11+18X12+18X13+20X14+20X15+22X16+18X21+20X22+20X23+21X24
+22X25+23X26+23X27+25X28+8X31+9X32+10X33+11X34+11X35+13X36
-20X12-24X13-44X15-24X16-20X21-20X22-16X23-4X24-16X25-4X26-4X31
-4X32-4X33-4X34-20X35-20X36>=0
16X11+18X12+18X13+20X14+20X15+22X16+18X21+20X22+20X23+21X24
+22X25+23X26+23X27+25X28+8X31+9X32+10X33+11X34+11X35+13X36
-28X11-28X12-28X13-56X14-28X15-56X16-12X23-28X24-12X25-28X26
-40X27-40X28-8X31-4X32-4X33-4X35>=0
16X11+18X12+18X13+20X14+20X15+22X16+18X21+20X22+20X23+21X24
+22X25+23X26+23X27+25X28+8X31+9X32+10X33+11X34+11X35+13X36-8X11
-8X12-8X13-8X14-8X15-8X16-40X21-40X23-40X24-40X27-8X32-12X32
-20X34-12X36>=0
END
```

MOLP model without workload balancing (P2) :

```
MIN. M+0.3333333C1+0.3333333C2+0.3333333C3-10
S.T.
X11+X12+X13+X14+X15+X16>=8
X21+X22+X23+X24+X25+X26+X27+X28>=6
X31+X32+X33+X34+X35+X36>=12
7X11+4X12+3X13+4X14+3X21+15X22+3X23+3X24+15X25+15X26+3X27+15X28
+5X31+5X32+5X33+5X34+5X35+5X36-M<=0
5X12+6X13+11X15+6X16+5X21+5X22+4X23+1X24+4X25+X26+X31+X32+X33
+X34+5X35+5X36-M<=0
7X11+7X12+7X13+14X14+7X15+14X16+3X23+7X24+3X25+7X26+10X27+10X28
+2X31+X32+X33+X35-M<=0
2X11+2X12+2X13+2X14+2X15+2X16+10X21+10X23+10X24+10X27+2X32+3X33
+5X34+3X36-M<=0
18.0X11+20.2X12+20.2X13+21.3X14+22.2X15+23.8X16-C1<=-10
19.8X21+21.6X22+21.8X23+23.0X24+24.0X25+24.8X26+25.0X27+27.2X28
-C2<=-20
9.6X31+10.8X32+11.8X33+12.6X34+13.0X35+15.0X36-C3<=-30
C1<=200
C2<=183
C3<=210
C1-M<=0
C2-M<=0
C3-M<=0
END
```

Step 1. Results of the two MOLP models (P1, P2)

P1 (MOLP with workload balancing)

Production cycle	Part type	Number of parts routed according to alternative routes								Sum	Total
		1	2	3	4	5	6	7	8		
1	A	0	0	5	0	3	0	0	0	8	
	B	3	0	0	3	0	0	0	0	6	26
	C	9	0	0	3	0	0	0	0	12	
2	A	0	0	10	0	6	0	0	0	16	
	B	6	0	0	6	0	0	0	0	12	52
	C	17	0	0	7	0	0	0	0	24	
3	A	0	0	20	0	12	0	0	0	32	
	B	13	0	0	11	0	0	0	0	24	104
	C	34	0	0	14	0	0	0	0	48	
4	A	0	0	38	0	22	0	0	0	60	
	B	23	0	0	22	0	0	0	0	45	195
	C	64	0	0	26	0	0	0	0	90	
5	A	0	0	51	0	29	0	0	0	80	
	B	31	0	0	29	0	0	0	0	60	260
	C	86	0	0	34	0	0	0	0	120	
6	A	0	0	63	0	37	0	0	0	100	
	B	39	0	0	36	0	0	0	0	75	325
	C	107	0	0	43	0	0	0	0	150	

P2 (MOLP without workload balancing)

Production cycle	Part type	Number of parts routed according to alternative routes								Sum	Total
		1	2	3	4	5	6	7	8		
1	A	8	0	0	0	0	0	0	0	8	
	B	6	0	0	0	0	0	0	0	6	26
	C	12	0	0	0	0	0	0	0	12	

```
---------------------------------------------------------------
       A    16   0   0   0   0   0   0   0    16
  2    B    12   0   0   0   0   0   0   0    12     52
       C    24   0   0   0   0   0   0   0    24

       A    32   0   0   0   0   0   0   0    32
  3    B    24   0   0   0   0   0   0   0    24    104
       C    48   0   0   0   0   0   0   0    48

       A    60   0   0   0   0   0   0   0    60
  4    B    45   0   0   0   0   0   0   0    45    195
       C    90   0   0   0   0   0   0   0    90

       A    80   0   0   0   0   0   0   0    80
  5    B    60   0   0   0   0   0   0   0    60    260
       C   120   0   0   0   0   0   0   0   120

       A   100   0   0   0   0   0   0   0   100
  6    B    75   0   0   0   0   0   0   0    75    325
       C   150   0   0   0   0   0   0   0   150
===============================================================
```

Step 2. Results of the CAN_Q implementation

```
===============================================================
      Performance measures      P1                 P2
===============================================================
      Production rate     7.886 parts/hour   7.732 parts/hour
q=3   Mean flow time       220.64 minutes     225.04 minutes
      System utilization        68 %               62 %
---------------------------------------------------------------
      Production rate     7.895 parts/hour   7.832 parts/hour
q=5   Mean flow time       342.00 minutes     344.76 minutes
      System utilization        69 %               63 %
===============================================================
```

Step 3. We choose the solution of P1 (MOLP with workload

        balancing) based on the results of step 2.

D. Simulation Experiments

   A simulation model using SLAM II and FORTRAN is developed. The

simulation experiments are conducted to compare the five process selection rules at six different levels of outputs, 26, 52, 104, 195, 260, and 325, respectively. Since data elements are assumed to be known in advance, each of the 120 tests (6 output levels x 5 process selection rules x 2 buffer sizes x 2 number of AGVs) takes a single replication. The multi-criteria performance measures considered simultaneously in each simulation are makespan, mean flowtime, mean tardiness, maximum tardiness, and system utilization for all machines.

IV. SIMULATION RESULTS AND VALIDATION OF THE MODEL

Fig. 3 through Fig. 12 show the results of the simulation experiments in terms of two different capacity levels (q=3, q=5) for the five kinds of performance measures.

1) The overall comparison of the seven process selection rules and methods for all performance measures is shown as follows :

    q = 3 : NAR (P2) < ARP < MARP (P1) << WINQ < ARPD < ARD = MARPD

    q = 5 : NAR (P2) < ARP << MARP (P1) < WINQ < ARPD < ARD < MARPD

2) MARPD gives the best results for all cases.

3) The difference between the group of static methods (MARP, ARP, and NAR) and the group of dynamic methods (MARPD, ARD, ARPD, and WINQ) are significant in every performance

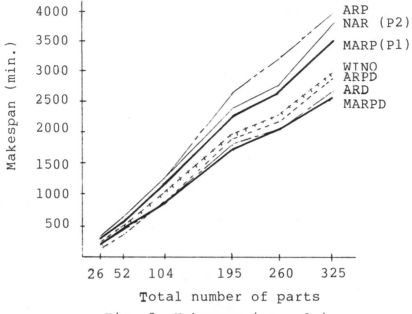

Fig. 3. Makespan (q = 3)

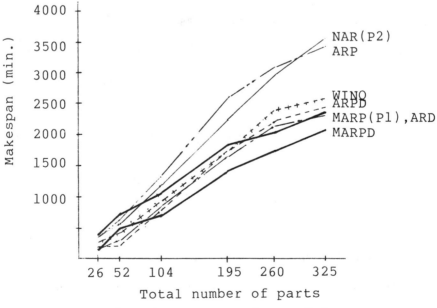

Fig. 4. Makespan (q = 5)

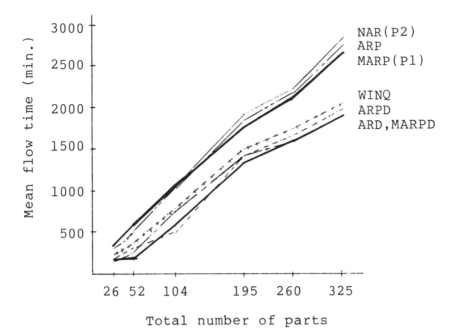

Fig. 5. Mean flow time (q = 3)

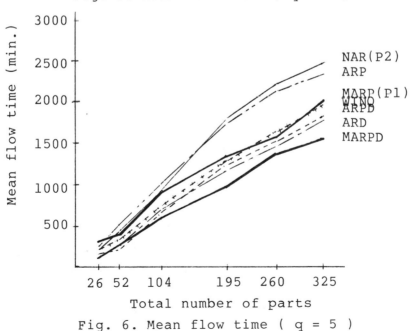

Fig. 6. Mean flow time (q = 5)

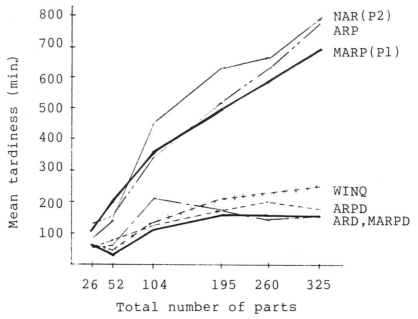

Fig. 7. Mean tardiness (q = 3)

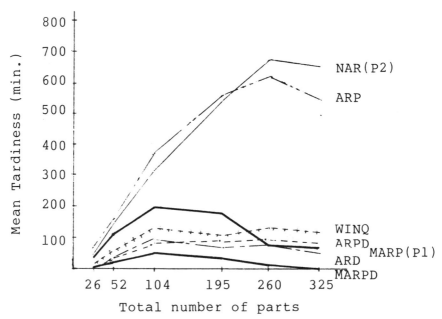

Fig. 8. Mean Tardiness (q = 5)

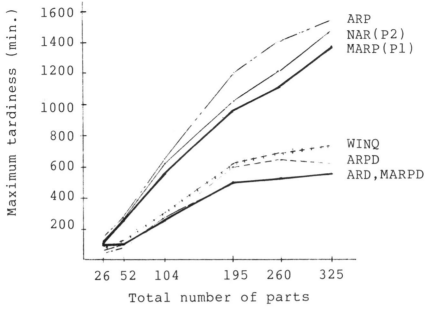

Fig. 9. Maximum tardiness (q = 3)

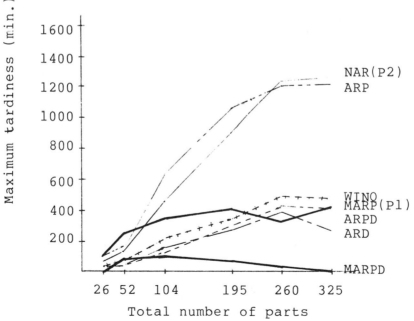

Fig. 10. Maximum tardiness (q = 5)

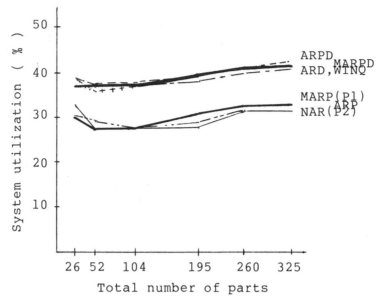

Fig. 11. System utilization (q = 3)

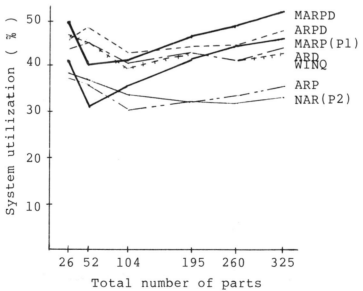

Fig. 12. System utilization (q = 5)

measure. Additionally, this differences are more significant as the production requirements are increased. Exceptionally, in the case of q = 5, MARP which is one of the static methods shows the similar results with the dynamic methods as the production requirements are increased over 195.

4) There are little differences within the groups of static and dynamic methods, respectively. However, when q = 5, MARP and MARPD give much better results than the others within the groups of static and dynamic methods, respectively.

5) The case of q = 5 shows the better results than the case of q = 3 because machine idle times are reduced under the assumption that the machine capacities are infinite.

6) In the mean tardiness and maximum tardiness, MARPD gives the value of zero (negative lateness) as the production requirements goes over 325 when q = 5.

The validation procedure of the simulation model in this paper follows that of the previous paper [19].

Rather than proving that a simulation model is a true model of a real system, simulation practitioners talk of having levels of confidence. In practice, it is useful to distinguish between verification and validation. According to Law and Kelton [20]

" Verification is determining whether a simulation model performs as intened, i.e., debugging the computer program. Validation is determining whether a simulation model is an accurate representation of the real world system under study ".

Since the simulation model described in this paper is based on a hypothetical FMS, only verification is necessary and it involved the following stages :

1) Debugging of the SLAM II and FORTRAN code

2) Checking the internal logic of the model

3) Comparing the model output with information gathered from a
   manual simulation using the same data.

The first three steps are facilitated by the TRACE option in SLAM II. The TRACE report generates a detailed account of the progress of a simulation. Verification involves checking the output report line-by-line so as to ensure that the model accurately reflected the hypothetical FMS.

V. CONCLUSIONS

In this paper, the scheduling and operational control problems for jobs and AGVs in an FMS having local buffers with limited capacities are discussed. In the process selection problems, the overall comparison of the existing and new rules and methods for all performance measures is as follows :

q = 3 : NAR < ARP < MARP << WINQ < ARPD < ARD = MARPD

q = 5 : NAR < ARP << MARP < WINQ < ARPD < ARD < MARPD

As shown in the above, MARPD gives the best results for all cases. That is, the method considering multiple objectives and the combination of the several techniques (such as MOLP, closed queueing network, and dynamic control method) gives better results than the existing rules and methods for all performance measures. In addition, the case of $q = 5$ shows the better results than the case of $q = 3$ because machine idle times are reduced under the assumption that the machine capacities are infinite.

VI. REFERENCES

1. S. Khator and S.Y. Nof, " A Flow Control Modelling Approach for Computerized Manufacturing Systems," Material Flow, Vol.2, 1-10, 1985.

2. S.Y. Nof, M.M. Barash, and J.J. Solberg, " Operational Control of Item Flow in Versatile Manufacturing Systems," INT. J. PROD. RES., Vol.17, 479-489, 1979.

3. I.K. Ro and J.I. Kim, " Multi-Criteria Operational Control Rules in Flexible Manufacturing Systems," INT. J. PROD RES., Vol.28, 47-63, 1990.

4. J. Kimemia and S.B. Gershwin, " Flow Optimization in Flexible Manufacturing Systems," INT. J. PROD. RES., Vol.23, 81-96, 1985.

5. W.E. Wilhelm and H.Y. Shin, " Effectiveness of Alternate operations in A Flexible Manufacturing System," INT. J. PROD. RES., Vol.23, 65-79, 1985.

6. A.J. Van Looveren, L.F. Gelders, and L.N. Wassenhove, " A Review of FMS Planning Models," Modelling and Design of Flexible Manufacturing Systems (A. Kusiak ed.), Elsevier, New York, 3-31,1986.

7. L.H. Avnots And L.N. Van Wassenhove, " The Part Mix and Routing Mix Problem in FMS : A Coupling between an LP Model and a Closed Queueing Network," INI. J. PROD. RES., Vol.26, 1981-1902, 1988.

8. S-Y. D. Wu and R.A. Wysk, " An Application of Discrete-Event Simulation to On-Line Control and Scheduling in Flexible Manufacturing," INT. J. PROD. RES., Vol.27, 1603-1623, 1989.

9. S.M. Lee and H.J. Jung, " A Multi-Objective Production Planning Model in a Flexible Manufacturing Environment," INT. J. PROD. RES., Vol.27, 1981-1992, 1989.

10. R.H. Conway, W.L. Maxwell, and L.W. Miller, " Theory of Scheduling," Addison-Wesley, 1967.

11. K.E. Stecke and T.L. Morin," The Optimality of Balancing Workloads in Certain Types of Flexible Manufacturing Systems," European Journal of Operational Research, Vol.20, 18-82, 1985.

12. J.P. Ignizio, " Linear Programming in Single & Multiple Objective Systems," Prentice-Hall, Englewood Cliffs, N.J., 1982.

13. J.J. Solberg, " Optimal Design and Control of Computerized Manufacturing Systems," Proceedings, AIIE Systems Engineering Conference, Boston Massachusetts, 1976.

14. J.J. Solberg, " A Mathematical Model of Computerized Manufacturing Systems," Proceedings, 4th International Conference on Production Research, Tokyo, 1977.

15. H.C. Co and R.A. Wysk, " The Robustness of CAN_Q in Modelling Automated Manufacturing Systems," INT. J. PROD.

RES., Vol.24,1485-1503, 1986.

16. J.A. Shantikumar and J.E. Steck, " Reducing in Work-In-Process Inventory in Certain Classes of Flexible Manufacturing Systems," European Journal of Operational Research, Vol.26, 266-271, 1986.

17. K.R. Baker, " Sequencing Rules and Due Date Assignments in a Job Shop," Management Science, Vol.30, 1093-1104, 1984.

18. N. Raman, B. Talbot, and R. Rachamadugu, " Simultaneous Scheduling of Material Handling Devices in Automated Manufacturing," Proceedings of the Second ORSA/TIMS Conference on FMS : Operations Research Models and Applications (K.E. Steck and R. Suri ed.), Elsevier, Amsterdam, 455-465, 1983.

19. P. O'Gorman, J. Gibbons, and J. Brown, " Evalution of Scheduling Systems for a Flexible Transfer Line Using a Simulation Model," Flexible Manufacturing Systems : Methods & Studies (A. Kusiak ed.) North-Holland, Amsterdam, 209-222, 1986.

20. A.M. Law and W.D. Kelton, Simulation Modelling and Analysis, McGraw Hill, New York, 1982.

COMPUTER–AIDED MANUFACTURING SYSTEMS DESIGN: FRAMEWORK AND TOOLS

Bin Wu

Department of Manufacturing and Engineering Systems
Brunel University, England

I. INTRODUCTION

The aim of this text is to present a framework of computer–aided manufacturing systems design (CAMSD), and discuss the tools and methodologies which can be used in an integrated environment.

A. THE NEED FOR AN EFFECTIVE FRAMEWORK

The significance of an effective approach for the design and evaluation of modern manufacturing systems is reflected by some of the common problems faced by today's manufacturing industries. These include:

> *Frequent technological and organizational changes.* Today's manufacturing industries must become more competitive because of the increasing international competition and customer demand for more variety, higher quality and lower price. This often requires a faster new–product–to–market cycle. Advanced manufacturing technologies (AMT) must be continuously introduced into their operation in order to provide the necessary flexibility and productivity. Therefore, either pulled by

market demands or pushed by new technology development, firms are having to restructure more frequently than ever before.

Need for accurate initial specifications. AMTs' high level of sophistication has also created the need for a high quality level of systems design and implementation. This is because the costs of failing to implement a system correctly is much greater today than previously, due to the higher level of capital investment involved. The system components require high levels of interaction with each other to bring about the desired transformations, and as there is less and less human intuition in their operation to deal with ambiguities, careful planning of such relationships is becoming increasingly critical.

Therefore, today's firms often need to restructure their operations in order to satisfy new demands, but in the same time the traditional trial–and–error approach to gradually reach a steady state of systems operation is no longer feasible. This has led to the realisation that better methods had to be used so that the chances of mistake in systems design are reduced, and the systems effectiveness are enhanced when put into operation. It has been manifested by today's most successful manufacturing organizations that there is more to a firm's success than simply running all its machines at maximum capacity, without considering the performance of other parts of the organisation or the impact of the environment. In general, when used in the past, the traditional local–optimizing approach has not effectively improved the overall performance of a system. When viewed from a wider perspective, manufacturing operations exhibit many systems characteristics and should therefore be tackled accordingly. This is the major reason why it is felt nowadays that a systems approach should be viewed as a more adequate framework for the analysis of problems which are generated by modern manufacturing operations. The most effective manufacturing operations always have clearly defined objectives. While the setting of these objectives can be a difficult and unstructured problem, the actual process of systems design which must create a system capable of fulfilling these objectives, is often a relatively well structured problem which is suited for a "hard" systems approach [1]. Such an approach will seek to make a purpose–oriented manufacturing organization as simple as possible by dividing the system as a whole into sub–systems, by clearly identifying the inputs, processes and outputs of each sub–system, by determining the appropriate controls to monitor achievement against predetermined standards, and by formulating and initiating the necessary corrective action. The division of the system into sub–systems during a design process breaks the complexity which is inherent in manufacturing systems into controllable units which can then be managed through a coordinated manner. Such a design approach should create a system in which the effective flow and interaction of information, materials, personnel, equipment, money and so on, are achieved to guarantee a smooth operation in order to effectively support a set of certain business goals. The framework will serve as a guide to the systems designer, reducing the amount of effort expended in deciding what to do, and ensuring that all aspects which require consideration are given that consideration so that a better-quality solution is produced. Although the methodology cannot offer solutions for specific problems, it can suggest tools which can be used in certain situations. It must also help ensure that innovative ideas are supported.

B. THE NEED FOR A COMPUTER-INTEGRATED APPROACH

The design of a manufacturing system is a complex task because of the very nature of manufacturing operations. For even a small organization the information needed will be quite considerable. It is obvious that the successful use of a design framework in practice will need the information–handling and analytical power of computer–aided tools. Up to date there have been a number of such tools available to help the systems designer in modelling and analyzing manufacturing systems. These include computer–aided system specification (such as IDEF, i.e., Integrated computer–aided manufacturing DEFination), computer simulation, computer–aided optimization and expert systems. Although potentially flexible and powerful, the majority of their applications have been in the hand of experts (despite the recent effort to utilize knowledge–based models of optimization). Furthermore, up to date they are still being used mainly for individual tasks in an isolated manner.

The way in which a computer modelling and simulation project is typically carried out clearly illustrates these characteristics (Figure 1). The first issue is the data availability for the construction of a simulation model and then its experimentation. Simulation models mimic a real system and they therefore must be provided with "pseudo–data" that details the conceived system, such as processing and demand information, and operation rules. Data collection is regarded by many as one of the most costly and time consuming tasks. Even under the best circumstance a simulation project frequently takes up to many months before useful results are realised.

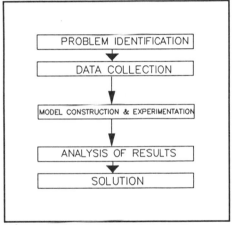

Figure 1
Simulation modelling

In addition, simulation experiments are of a statistical nature and so the results have to be treated as such. This requires a sufficient number of simulation runs for each system configuration. Without the help of a computerized information handling system, the number of design alternatives that can be simulated in a reasonable time is limited, hence restricting the designer's freedom to generate more alternative designs. These observations are also generally true for the applications of analytical (mathematical) modelling and expert systems. That is, they all follow a similar problem–solving pattern as shown in Figure 1 and can suffer from a similar set of shortcomings. Furthermore, the way in which they are being applied (i.e., independently to each other) can result in wasted effort in some of the common areas. These problems may be solved (or at least the situation can be improved) by integrating the individual tools through a global CAMSD database, as shown in Figure 2. For example, through this database simulation and expert system can be used as complementary technologies that are combined to provide a more powerful decision technique than either of the two on its own. A similar integration can also be achieved between simulation and certain analytical models, for example, with the simulation model generating cost estimation which are than used by a cell formation model to produce optimum cellular configuration [2]. Therefore, amongst other things such a CAMSD database will serve the following two purposes:

a. *As a formal information management system to support the design framework.* As will be seen later, most of the methodologies developed to date break the total design process into a set of broad phases, and these project phases are then broken down into smaller, more manageable steps. Although iteration is always a necessity, these methodologies usually require a set of fixed outputs from one stage before continuing to the next. For instance, a "conceptual design phase" may yield a conceptual model of the system under consideration in terms of types of product, system inputs and overall capacity required, whilst a "detailed design phase" will produce a detailed list of the parts to be produced, the production processes to be followed, and the number and type of machines to be used. All of these need to be accurately recorded and stored so that they can be easily retrieved for updating, modification, or to be used as inputs to the subsequent design tasks. With a CAMSD database designed according to the information structure of the design framework, the information recording and retrieval processes can be made much more efficient and less tedious.

b. *As a platform to link various computer–aided tools within an integrated CAMSD environment.* Within an integrate environment through such a database, the necessary data for any specific task of analysis and evaluation can be more easily identified and obtained, thus reducing the duplicated effort in data collection and analysis for individual tasks. Also, the various computer–aided tools can be utilized in a more coherent manner rather than in the stand–alone approach of today.

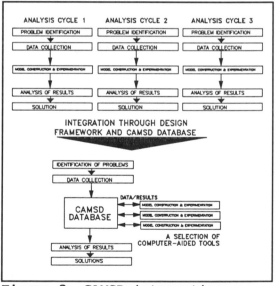

Figure 2   CAMSD integration

The discussion in the remainder of this text will mainly follow a general design framework proposed previously by this author (B.Wu, *"Fundamentals of Manufacturing Systems Design and Analysis,"* Chapman and Hall, London, 1991). Section II will review the main steps of this design framework, together with an analysis of the structure and requirements of a CAMSD database developed to support it. A review of some of the analytical methods and computer–aided tools, which can be used within the design framework to provide integrated CAMSD, is given in Section III.

II. STRUCTURE OF THE DESIGN FRAMEWORK

Two distinct approaches can be taken to system design, i.e., the top–down and the bottow–up. The former starts with a set of objectives and then creates a system model which fits the intended purpose. This approach gives little consideration of the current system being operated. Although preferable from the designer's point of view, it can result in a design which requires a total replacement of the current system and therefore heavy capital investment. The other approach bases its consideration mostly on the existing system, producing designs which require less capital investment. However, since it is unlikely that any manufacturing unit was initially designed to cope with later structural and functional alternations, the options available could be constrained and ideas severely limited. The approach outlined here is a hybrid of these two approaches, which initiates a project by analyzing the current and the desired future positions of a system of concern. The current position of the system will always be given first consideration. Analysis of the desired future position will then identify a set of new objectives, which will be under the constraint of the findings from the previous analysis. A new system model can then be developed which will fulfil these objectives. The creative thought for the new system is not therefore confined by the workings of the old system – only the objectives are constrained by the realistic starting position adopted.

The structure of the design process is based upon a general model of problem-solving cycle, as shown in Figure 3. The first two stages of the general model are the "analysis of situation" and "setting of objectives". Both of them fall into a field of study which could be loosely defined as "manufacturing strategy", which first requires an analysis of the current state of operation of a manufacturing organisation under consideration, and then initiates an analysis of its current markets and their future prospects, and the prospects of potential markets which could be entered. Through such exercises, a vision is formed of where the firm should be in a future situation and what should be done in order to achieved that state. The following phases

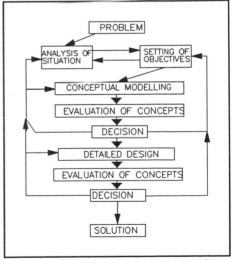

Figure 3 Framework overview

identified in Figure 3 are the design phases which actually produce and evaluate plans that will be the basis for transforming this operation from its current to the desired state. These will produce a conceptual model which defines the overall structure of the new system to achieve the desired results, and then fill in the details to form a complete design. The individual stages of the design methodology as shown in Figure 3 will now be described in more detail in the following text.

A. ANALYSIS OF SITUATION

In a manufacturing environment there exist a number of possible causes for change. The first type of changes can be explicitly known to the operational managers if they are objectives issued from their wider system (e.g., from higher level management). The second reason for change is due to a perceived change in the environment which will give the firm an opportunity for innovation, for instance, through a marketing exercise or product development function which usually highlights future requirements. The third is due to unsatisfactory performance of an existing system, e.g., slower throughput times or poor profit figures. Whatever the signs of impending change may be, it is always necessary to establish exactly what the problems are. A preliminary investigation should therefore be carried out which will produce a *Symptoms List* containing all those items which are causing concern, and then identify the roots of the problems. This process is rather like planning for a journey. To reach B from A, for instance, the starting point and end point of the journey must be known in advance if the best route is to be selected. Similarly, to accomplish the best system changes, both the starting point and the desired finish should be known. This requires the understanding of the current system, its components and their interaction. It is then necessary to understand how the current system fails to achieve the current or future requirements by identifying the reasons for the problems (In general, the three areas in which analysis of the current position must be carried out are marketing, product, and manufacturing system. The first two are beyond the scope of this design frame. Nevertheless, their inputs are required and must be gained from the necessary sources). In order to reduce the possibility of problem areas being hidden due to bias, a structured procedure for system analysis and identification should be adopted – such as the "What–Which–What" approach outlined below.

1. WHAT – finding what are involved in the system

1.a   **Analysis of current production technology and personnel skill.** The first aspect to be considered is the production technology. All of the currently used production technology should be recorded in a *Plant List*, together with a map of the current layout of all equipment. The functional range of each item of plant should be recorded in terms of its ability and capacity, e.g.:

> type of operation
> size of components which can be processed
> number of different tools which can be loaded
> limits on accuracy which can be achieved
> availability of the machine in standard hours

Once completed, the *Plant List* will provide a complete, static picture of the current system capacities within the company. In addition, a *Subcontract*

Function List should also be created to record manufacturing functions which are available to the system, but are outside the organisation with the subcontractors used recorded in a separate *Subcontractor List*. Finally a list of personnel infrastructure will need to be created, which contains all of the manpower and skills provided by the company employees.

1.b   **Engineering analysis of products.** Next, a *Current Product List* should be created. This should identify the type and quantities of finished parts currently being produced. For each product listed in this list a *Part List* should also be produced. This allows a *Current Part List* to be generated which specifies every part which must be manufactured or procured, together with the demand levels.

1.c   **Process analysis of production.** Identification of process plans for each of the parts recorded in the *Current Part List* will then provide indication about the type and capacity of manufacturing facilities required within the manufacturing system. These functions should be recorded in a *Current Manufacturing Function List*, and cross referenced to the *Current Part List*.

A large amount of these information can be obtained by directly accessing the company's current manufacturing information system (such as a MRP II system). Once completed, these lists can be used later for a number of purposes.

2. WHICH AND WHAT – identifying elements of interest

By identifying "which" areas are of interest and "what" these are, it is possible to conceptually arrange them as components and/or subsystems in order to aid the subsequent investigation:–

2.a   **Systems description.** The manufacturing functions at this stage may be modelled and described using either input–output models or the $IDEF_0$ technique. The resultant system model should represent both the functional and information structure of the manufacturing operation as accurately as possible at an appropriate level of data aggregation. The analyst can then assess the list of problems and where they are occurring in the system model. The departments identified will need further decomposition to allow full assessment of the problem.

2.b   **Static and dynamic analysis.** To identify the "Which" and "What", it is necessary to consider the resultant model of system description in relation to the *Symptom List*. For example, when new products are to be produced, the concern will be in terms of new functions or increased capacity required from existing functions.   All of these items should be identified and added to a *Problem/Requirement List*. In addition, problems associated with the dynamic characteristics of the system should also be analyzed. There are a number of useful tools for dynamic investigation, and one of the favourite of these is discrete computer simulation. Such dynamic analysis, when completed, should reveal the underlying problems causing the remaining symptoms.   These should again be added to the *Problem/Requirement List.*

B. SETTING OBJECTIVES

Inputs		Process		Outputs
Cost analysis	—>		—>	Trade-offs resulting from strategy options
Inventory analysis	—>		—>	
	—>		—>	Reconciliation of manufacturing capability and
Purchasing analysis	—>		—>	mkt opportunities
Product analysis	—>	Manufac- turing	—>	Make or buy decisions
	—>	Strategy	—>	
Production planning analysis	—>	Develop- ment	—>	Process choices
	—>		—>	Inventory levels - function and
Engineering objectives	—>		—>	position as part of the total strategy
Business objectives	—>		—>	investments to reflect the
	—>		—>	manufacturing trade-offs to
Marketing/ Product strategy	—>		—>	meet the market needs
-product range	—>		—>	
-design -sales forecasts	—>		—>	Infrastructure -organisational style
-competitive position of	—>		—>	-controls -work structure
each product	—>		—>	-procedures
-order winning criteria	—>			
R & D strategy	—>			

Figure 4 Manufacturing strategy development

Having identified the problem areas through the previous analysis phase, the resultant document will be used as the basis for the setting of realistic objectives which should in general adhere to the company's overall manufacturing strategy. This stage of the design process is closely related to manufacturing strategy. In most cases, the system which is being designed or redesigned will be part of a wider system, such that there will be a hierarchy of strategies which will ultimately lead back to the decisions and strategies adopted by the corporate board. Therefore, the system's specific objectives must integrate with the aims of other parts of the enterprise. It is nowadays a widely accepted view that successful firms must concentrate on a few aspects of their performance in order to achieve a competitive advantage. To achieve this, there should be a knowledge throughout a firm of what must be achieved and the priorities attached to these. For example, the five competitive stances identified by Slack [3] which a firm must balance are:–

Quality advantage,
Cost advantage ,
Response advantage,
Dependability advantage, and
Flexibility advantage.

The defining of the relative importance of these should be carried out by the higher level of company management so as to provide the necessary business background for the design process. Given in Figure 4 is an example for the analysis of manufacturing strategy development, showing the possible inputs and outputs of the process.

The results from this stage should be collected into a formal statement of objectives. This will indicate the priority of each of the objectives laid out. Where appropriate, it should also include the nominal levels and threshold levels which must be achieved if an option is to be an acceptable solution to the problem. The outputs from this phase will therefore indicate a gap between the current and the desired future situations. The following design phases will follow the direction such identified and develop a solution which must bridge this gap.

C. CONCEPTUAL MODELLING

This step is necessary because although there exist certain similarities between many firms, particularly at higher levels of organisation, at the operational levels they disappear rapidly. A conceptual model is therefore needed to identify the building blocks of the particular system concerned. These blocks will be in terms of a combination of manufacturing functions, their relationships, and together with the necessary controlling functions, as shown in Figure 5.

1. Identification of functional requirements

As the technology in the product will usually dictate the type of manufacturing technology choices, the product range in question will have a major influence upon the system to be designed. In general, the desired product range will include new products, enhanced products and different quantities of current products. Each relevant part should

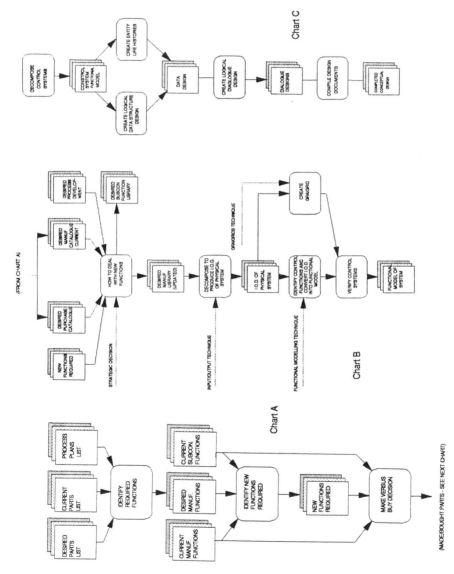

Figure 5 Conceptual modelling flow charts

be identified and recorded in a *Desired Part List*. The current products which are also in the desired product range can be copied from the *Current Part List* generated during the analysis phase. Estimates of demands should also be obtained and recorded in the list.

Next, the manufacturing functions which must be included in the new system should be identified. These are to be recorded in a *Desired Manufacturing Function List*. By comparing the current subcontract and manufacturing lists to the desired manufacturing list the manufacturing functions which are not currently available can be identified. Before further action can be taken it is necessary to decide how the expertise and capacity to support these functions should be provided. There will be three options available: acquiring expertise from outside, developing expertise in–house or simply subcontracting the work. The choice of make versus buy is a major strategic decision involved in manufacturing systems design. Shown in Figure 6 is an input/output outline of this decision process.

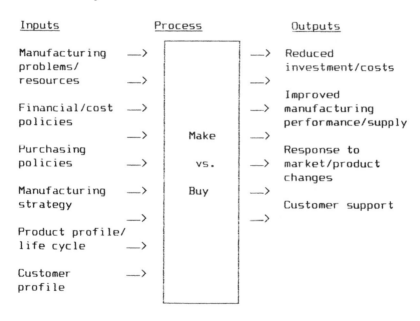

Figure 6 The make vs. buy process

The make vs. buy analysis of the *Desired Part List* will identified some parts which are to be made in–house, and others to be manufactured by subcontractors. The actual in–house capacity of manufacturing functions presently available have been identified in the *Current Manufacturing Function List*. The maximum required capacity for any function is the sum of the capacity required to satisfy the demands of the in–house parts already identified plus that needed for all of the undecided parts. The excess demand can be fulfilled either by subcontracting the work, or by increasing capacity of the factory.

2. Organisation of functions

The identified functions should be organised in such a way so that the objectives laid down can be effectively fulfilled. This task can be achieved using a decomposition tool which will allow a specific function of the system to be examined in detail, and in the same time still keep an overall perspective. The tools available for this task include various graphical methods, with $IDEF_0$ and input–output diagrams (IOD) being the most frequently used.

The process involved here in describing the physical system will be similar to that of the manufacturing system model discussed earlier. For instance, the top layer of an IOD model will present the system operation as a single function with identified inputs and outputs. A description of the function will identify the purpose of the system, and the competitive stance to be taken. The inputs identified on the model will be all the materials and parts which are bought in for the manufacturing process. The outputs will include a summary of the information as given in the *Desired Part List*. Next the basic components identified for the system must be organised to produce the outputs of this IOD model. This is achieved by decomposing the model until the level concerning component manufacturing function is reached. It should be remembered that a number of different groupings can be produced dependent on the criteria applied to the decomposing process. To deal with this problem, alternative business system options (BSO) should be created which are system models with different organisational structure and technologies, but all aim to fulfil the same set of outputs. A considerable body of knowledge can be drawn from the manufacturing strategy literature in this regard. For example, AMT is more likely to be implemented successfully if systems are simplified. This is supported in such theories as the "focused factory" or the "plant within a plant" (e.g., see Skinner [4]). The criteria that can be used in the decomposition of the top level IOD include "decomposition by product/process matrix", "decomposition by competitive characteristics" and "decomposition by process", etc. From a number of such BSOs, the final selection of a BSO must be taken on a strategic basis for further development, which will be used as the skeleton for the hierarchy of departments, cells and workstations, etc., of the final system.

3. Analysis of control systems

Each subsystem defined in the IOD will require its own internal control system sorted out, which is required to improve the effectiveness of the manufacturing operations concerned. Different areas of the organization will have different requirements on the type of control needed. Some areas will change in a dynamic manner either due to variation in the inputs or due to disturbance from the wider system. For this type of operation it would be desirable to have a continuous on–line control function which responds immediately to changes. On the other hand, there are also areas where things can be done in a deterministic (or near–deterministic) manner, permitting decisions to be made in an off–line, one per day or even per week basis. Important considerations here include the levels of synchronisation required, the amount of information which the systems have to process and the time which the processing takes. Wherever appropriate, smaller, distributed control systems should be used so that each level in the hierarchy only needs to deal with a small set of information. To achieve satisfactory operation, it is also essential that the different control systems are effectively coordinated. It should be pointed out that within a manufacturing environment, the control systems need not always to be computer based. Kanban card method, for example, is a well–known, non–computerised control approach. The representation of the required controlling functions on the physical system model can be achieved with a control function superimposed on each level of the decomposition, and the span of control being the manufacturing functions on this level. The following steps may be followed:–

a.   **Identify variables**. The variables which affect the functional output must be identified. It is important that the selected variable should have a relationship to a key objective and also be controllable.

b.   **Identify data flows**. The sources and destinations of the required data can then be identified and then added to the model. A functional description should be prepared which includes the decision horizons and decisions period, i.e., time for a decision to affect the output form the system.

c.   **Identify processes**. It is also necessary to understand how the variable can be used to control the functions in question. If a direct relationship is not known, it may be necessary first to build up knowledge prior to system construction, e.g., by using computer simulation modelling. Alternatively the control system can designed in such a way that knowledge may be accumulated in an on–line manner, e.g., by using such techniques as Statistical Process Control (SPC) or expert systems.

As a control system, whether manual or automated, is always based on an information system, the current methodologies developed for information systems design and analysis, such as SSADM [5], are sometimes useful to deal with certain design tasks involved. SSADM can be used, for example, to aid the data design process of a control system, helping clarify the information structure of the control functions and specify the data storage and management facilities.

The completion of the steps outlined in this section completes the conceptual modelling of the system. All documentation should be gathered and sorted for use in the detailed design phase.

D. DETAILED DESIGN

The completion of conceptual design of the system will have created a system model in terms of the related functions from the *Desired Manufacturing Function List*. It contains a set of interrelated manufacturing functions each with a related list of products, together with a hierarchy of control systems which will process a set of key information associated to the effective performance of these functions. In addition, the long term production capacity to be achieved will also have been specified in terms of the average static capacity levels and the levels of variation to reflect the dynamic requirements. The detailed design phase takes the conceptual model of the selected manufacturing system option and transforms this into detailed specifications which could be used later for implementation. Again two interrelated areas of consideration are involved, i.e., physical and control. Whilst the selected conceptual model dictates the type of functions and the relationships between them, detailed design must decide on the detailed layout of the plant by selecting and allocating the required manufacturing equipment, as well as the physical infrastructure including transportation and storage facilities. In the same time, the controlling functions must also be developed so as to bring about the desired outputs from the system. Some of the mathematical techniques and computer–aided tools that can be used during this stage of the design process will be discussed in Section III. The following text outlines the general tasks involved (see Figure 7).

1. Identification of cell requirements

First of all, it is necessary to decide on the configuration of the subsystems (or manufacturing cells) and provide detailed specification for the selection of most suitable items of plant. The approach to this tasks is illustrated in Figure 7.a. The functional requirements defined by the conceptual systems model act here as constraints on the solutions to be selected. As in general the equipment is not being purchased at present, a full specification and list of decision criteria should first be created and recorded with a *Workstation List* to provide accurate and detailed information. The items selected should be recorded in a *Cell List*. The most important factors to consider are those which allow the cell to fulfil the targets defined in the previous phase. They can be identified by utilising the experience of the design team, by obtaining external expertise and by discussion with future users.

An extremely important issue here is concerned with the financial appraisal of investment on advanced manufacturing technology. The implication of this subject is so significant that it must be regarded as an area of prerequisite knowledge for any system designer involved in today's manufacturing industry. Although AMTs are being increasingly adopted throughout the industry, they are becoming more complicated and expensive, both for stand alone machines and for systems. This has made acquisition of the necessary investment capital difficult. The problem is made even worse because many of the advantages claimed for AMT do not comply with the traditional defination of "productivity". The traditional methods of investment appraisal (e.g., payback or discounting cash flows) demand rapid repayment, whilst the new technologies are increasingly infra–structural, and provide long term and often intangible gains. An intangible gain is one which causes improvement in some aspect of the business and its operations, but the result is not clearly seen in financial terms. These include, for example:

* Ability to response to market change with respect to product volume, product mix and product change;
* Ability to response to specific customer requirement;
* Shorter lead time, reduced inventory, better product quality and consistency;
* Improved plant controls;
* Improved communication of information;

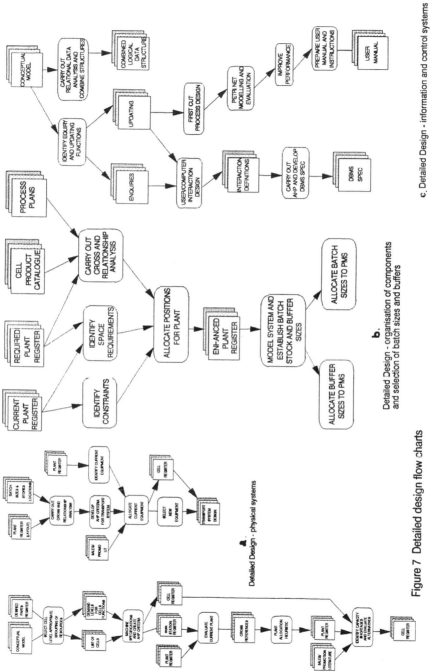

c. Detailed Design - information and control systems

b. Detailed Design - organisation of components and selection of batch sizes and buffers

a. Detailed Design - physical systems

Figure 7 Detailed design flow charts

*   Reduced tooling and simplified jigs;
*   Reduced overhead and direct labour content; and
*   Integrated manufacturing environment.

Therefore, there is a need for more complete methods of financial justification for AMT. Several approaches have been proposed that consider the problems of the vagueness and risks associated with intangibles [6, 7, 8]. For example, the analytical hierarchy process (AHP) takes a strategic approach [6], which initially sets an objective set for the required technology and then split this into sub–objectives, which must be fulfilled so as to achieve the initial objective. Through the involvement of a wide range of relevant personnel, intangible benefits are to be considered in relation to the set objectives. A figure is set for the level of fulfilment for each of these objectives. These can then be manipulated to allow a "utility vector" to be calculated for alternative choices. So far as AMT investment is concerned, this approach has a major advantage in that the intangible benefits, some of which may not even explicitly recognised by the design team, are more likely to be taken into consideration. This type of approaches should be adopted in the assessment of the suitability of new production facilities to fulfil the requirements of a cell (see Figure 7).

Obviously the current plant should be utilized to minimise the costs involved whenever possible. Each item of the plant should be considered for its possible application by assessing it against the hierarchy of selecting criteria. Upon the completion of this analysis, all of the machines which are capable of fulfilling a required function will be cross referenced to that function. After this process certain capacity requirements will still remain unfulfilled or partially fulfilled. It is then necessary to establish what new facilities should be used to fulfil them. This is one of the best opportunities to look for innovative options, since there are less constrains attached. In general the technology selected should be that which has lowest functionality whilst still fulfilling the operational and strategic objectives, as identified through the AHP process.

2. Layout of production facilities

Figure 7.b illustrates the processes involved in the layout of equipment and selection of buffer and storage facilities. The objectives of these processes should be in agreement with the overall objectives, and will often fall within three categories [9]:–

        to minimize cost of materials handling and movement,
        to minimize congestion and delay, and
        to maximize utilisation of space, facilities and labour.

The key word to remember here is again simplicity. It is particularly important to simplify material flows when advanced manufacturing system is concerned. This must be achieved within, as well as between cells. An advanced manufacturing system (AMS) is usually intended to increase flexibility. However, this does not mean that it must be possible to carry out any and every function, but instead tends to mean that it can achieve a range of this efficiently. When related to materials flow, this implies that in general provision can not be made for flows from any node to any other node. Careful selection of cell products should therefore attempt to identify a relatively wide range of products, but which have a limited amount of variability in their flow patterns (see Section III,A for analytical treatment of this subject). To achieve the best layout of workstations within cells, and the location of cells and departments in relation to one another, the space requirements of the functional groups should first be established. These can be estimated on the basis of expected floor space for each of the items of plant listed in the *Cell List*. One of the tools which can give guidance in the location of departments, as well as cells or workstations within cells, is the cross and relationship chart [9].

Storage space, batch sizes and buffer stocks can all have a major effect on the ability of the firm to meet its overall objectives. These must be constrained to achieve long term optimisation. For example, a firm's main strategy for a competitive advantage may not be the price, but offering prompt response for customized products or a higher dependability in delivery dates. Consequently, whilst in some cases it is desirable to keep the stock level down, in the others certain amount of safe stock may be necessary. A balance must then be achieved between the desired levels, and the costs which will incur. Once the batch quantities and the configuration of in process buffer stocks are established, it is then possible to consider what are the best forms of transportation. If transportation equipment is currently available, then they should be allocated to specific workstations, or to a general list of transportation equipment used between different cells, along with a map of the workstations or cells which it will serve.

To help establish the implications of different trade offs between batch size and in process storage quantities, a discrete event simulation model can again provide a useful tool. The level of aggregation of data is important here to reduce effort and expense involved in the creation of model and analysis of results. The input data can be either a set of known "typical" inputs, or can take the form of a randomly generated set of data, but which is known to possess the statistical characteristics of the true trends. The investigation of the system using a simulation model at this stage may also be beneficial later, for a similar model is likely to be used in the investigation of the control systems required. In fact, it is very likely that these two investigations will have to be undertaken simultaneously, as the control systems will not only effect the system performance, but also alter the way in which it will operate.

3. Development of control systems

The major tasks involved in the detailed design of the control system include the design of controlling processes and databases, the selection and location of computer hardware and the specification of managerial roles which will be responsible for certain decision centres. The procedure is illustrated in Figure 7.c. Two types of interaction will take place with a manufacturing engineering database. These are enquiries, which simply view the data, and updating processes which modify the data. The design of both these processes are dealt with adequately by SSADM. Having defined these interactions, it is then possible to proceed with specification of the actual processes. The final phase in SSADM (Stage 6) is specifically concerned with the design tasks in this regard. By following it, the first cut physical and process designs can be created. Their performance are then predicted to ensure that the processing fulfils the requirements. Upon completion of the performance appraisal, the basis of the detailed specification for the database management system are created together with file specifications.

It is necessary to ensure that the correct data processing devices are made available. This will require the selection of transducers, PCs and networking equipment, etc. For user interactions, the number of terminals or networked PCs and their locations, must also be established. The consideration of location of terminals should include identifying the location of manufacturing functions being controlled. Analytical techniques such as the operational research procedure of "travelling salesman" can be used to minimize the cabling requirements of a network. The physical locations and layout should be recorded in the *Plant List*.

In summary, therefore, there are three main areas to be considered in the detailed design stage of AMS. These are the selection of production facilities together with the selection of transportation and storage facilities, the configuration and layout of the facilities, and the detailed design of the control systems including both the hardware and software required. These processes are unlikely to occur sequentially, for decisions for each will have implications for the others. The output from this stage will be a design which is accurate to a high level, and detailed enough for actual implementation. However, these designs must still be assessed against the objectives which were originally set, so as to ensure that they are fulfilled and the investment involved is viable (see Figure 3).

E. EVALUATION AND DECISION

So far as manufacturing systems design is concerned, the major evaluating and decision-making activities usually take place following the conceptual modelling and the detailed design phases, as shown in Figure 3.

Following the conceptual modelling phase, it is necessary to assess each of the BSOs created with respect to each of the criteria which have been laid out in the objective setting phase. Most of the information needed will have to be generated through discussion within the design team, or by the use of external experts. In some cases it is possible to carry out rigorous assessment on the alternatives. For example, although the detailed designs of workstations have not been made available at this stage, it can still be feasible to utilise an aggregated simulation model to establish the acceptability of different groups within the system in terms of their ability to meet objectives. If a clear front runner can be identified, this will be automatically chosen. However, frequently a decision has to be based on several close options. Occasionally there may even be no acceptable solution at all. When this happens it is necessary to establish how easily a design can be modified to bring it to the acceptable level. The second evaluation and decision point is after the detailed design process. Evaluation of the systems future fulfilment of objectives can be achieved by using either computer simulation or other analytical means. Much of the proposed system will have already been modelled during the detailed design process, or at the higher levels during the earlier phases. However, at this stage a thorough evaluation of the systems is essential, and so the remaining subsystem must be included in the evaluation frame at a dis–aggregated level.

A realistic hurdle rate should be adopted for the decision–making process, which should be in line with the real returns required on a risk free investment, plus an allowance for the level of risk attached to the project. To assess the proposed system, opinion must be invited from a diverse range of sources. For example, discussion with marketing personnel may identify a possible increase in sales which could be expected from an improved level of quality. The total costs of the project should be estimated, along with a cash flow profile. Expenditures should not all be considered to occur at the beginning of the project, as distant prices will have a lower net present value, and thus the total costs will fall. The actual calculating process may be either carried out using a special purpose package, or with a spreadsheet based model. If the project fails on any of the criteria, it must be decided whether to continue with the development of the design, whether to modify the objectives, or whether to terminate the project. It is impossible to specifically define how to make a good decision in this regard, but the logic of this is straight forward. The additional cost to modify the design, and the expected change in cost for implementation of the system, should be added to the cash flow estimation. If the rate of return is achieved, then it is obviously desirable to continue with the project. If the rate of return is not achieved, but the project will still make positive inflows, the project should compared to the "do nothing" case. If however the project will clearly cost the firm more than it would earn, it must be terminated before further costs are incurred. However, the implications of those intangible benefits should always be bared in mind. The results of this analysis should be presented to the authorising body for a final decision.

F. THE STRUCTURE OF A CAMSD DATABASE

In practice, this framework can be made much more comprehensive and effective if the information involved are handled through a purpose built database. Such a database will be capable of satisfying the need for the storage and cross reference of information which are used first to analyze the existing system and then to model the new system. It will which take in, as well as provide with, the essential information as the design project progresses through each stage of analysis and decision–making. For instance, it should consist of a set of files, such as lists of the current and the desired manufacturing operations, which need to be compared and the differences copied to a list of new manufacturing operations. In general, this database should have the following characteristics:

* Allow relevant manufacturing data to be stored, updated and retrieved easily.

* Allow certain files to be compared and help identify the differences in terms of the types of entries and the magnitudes of their attributes.

* Allow specified sets of data to be automatically transferred to other parts of the database and ensure the necessary updates are made on all appropriate records.

*   Provide data input/output interfaces to facilitate the effective application of a
         range of computer–aided tools.

*   Comprehensive enough to provide appropriate information handling at every level
         of the design framework.

*   Flexible enough to be able to cope with the different requirements of a wide
         range of design situations.

Table I   Example of a data dictionary

NO	NAME	VIEWS	TYPE	WIDTH (dec)	SOURCE
1	PART NO	1,3,8,9,11	CHAR	12	B
2	PART NAME	1,2,8,9	CHAR	20	B
3	PARENT PART NO	1,2	CHAR	12	B
4	NO. NEXT ASSEMBLY	1,2	NUM	3(0)	B
5	CURRENT DEMAND	2	NUM	6(0)	A
6	DERIRED DEMAND	1	NUM	6(0)	B
7	S/C DEMAND	8	NUM	6(0)	B
8	I/H DEMAND	9	NUM	6(0)	C (VIEW 1)
9	NEW/OLD/ENHANCED	1	CHAR	1	B
10	OPERATION NO	1,3–7,11–13	CHAR	4	B
11	OPERATION TIME	1,3,11	NUM	6(2)	B or C (VIEW 3)
12	OPERATION NAME	1,3,11	CHAR	20	B or C (VIEW 3)
13	FUNCTION NO	4–7,12,13	CHAR	4	D or C (VIEW 5)
14	FUNCTION NAME	4–7,12,13	CHAR	20	D or D (VIEW 5)

The data requirements of individual views (a view in a database is a collection of data
which the user needs to view in order to make a particular judgement or analysis
decision) in relation to the design flow charts (Figures 5 and 7) allow a data dictionary
to be compiled, as shown in Table I, which is needed for the physical design and
implementation of the database. The relationships between these entities can also be
established by using entity relationship diagrams [10], see Figure 8. The source of each
data item is also indicated in Table I. The following four data sources are identified:

A.   ANALYSIS OF THE CURRENT SYSTEM. Data from this source describe the
        characteristics of the existing system. This type of information would are
        recorded during the initial analysis of the current manufacturing system.

B.   BRIEF FOR THE NEW SYSTEM. The brief of the new system is the list of parts which must be made within it, including the expected demand and other production

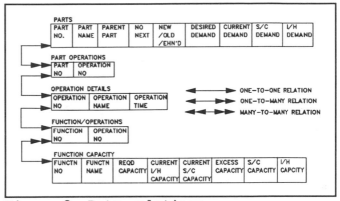

Figure 8   Data relations

data. This type of information comes from the initial analysis of market, and the design plans for how to make any new products in the future portfolio.

C.   COMPARISON OF VIEWS. The database itself must generate some of the information by comparing lists of existing information, or simply copying information from another view. Data items from source "C" make up the processing requirements for a particular view (i.e., what the database must be able to do with the data which it holds).

D.   DECISIONS MADE BY THE USER OR RESULTS FROM THE TOOLS. These are the results from the various decision-making processes, such as the make vs. buy analysis. They include the solutions generated by those computer-aided design tools.

III. ALGORITHMS AND COMPUTER-AIDED TOOLS

There are a wide range of modelling methods and tools which can be used to aid the various processes of evaluation and decision within the framework. These include:

*   Tools for objective identification, e.g., methods of systems analysis.

*   Tools for static analysis, e.g., spreedsheet based appraisal models for capital investment.

*   Tools for dynamic analysis, e.g., simulation, queuing models and petri-net models.

*   Tools for equipment selection, cellular configuration and facility layout, e.g., algorithms for enumeration, simulation and expert systems.

These can be loosely classified into three categories, as listed in Table II.

Table II Summary of Model Types

MODEL TYPE	ANALYTICAL	SIMULATION	EXPERT SYSTEM
Method of prediction	analytical	logical manipulation	reasoning
Method of evaluation	mathematical	experimentation	knowledge based
Cost	medium	high	high
Ease of communication	very poor	excellent	good
Weakness	only able to cope with simplified cases, specialists required	optimal solution not guaranteed, time consuming	difficult to extract knowledge, application usually restricted to a narrow area

A. ALGORITHMS FOR CELL CONFIGURATION

A wide range of algorithms have been developed for design tasks such as process planning, equipment selection, cellular configuration and facility layout. In the following text only those associated with cellular formation will be discussed to reveal the common characteristics and problems of these algorithms when put into practice, and to examine how they may be integrated into the CAMSD framework.

As the basis to develop FMS cells and also due to the need to simplify the control functions in an advanced manufacturing environment, cellular manufacturing is receiving increasing attention. In a conventional manufacturing layout the components pass through various functional sections which specialize in specific operations such as milling, drilling and deburring. Although the physical layout of this kind of organization is relatively simple, the logical layout (i.e., production flow) is usually messy and thus a complex control system is required to track each of the components around various sections in order to schedule them to maintain their appropriate priorities. In contract, a cellular based manufacturing system is arranged into a series of small manufacturing units known as cells. These are a small collection of machines and operators which produce a defined group of components. Due to the similarities between the components produced by such a cell, the production facilities within each cell are operated like a flow line, hence greatly reducing the amount of material handling required and the complexity of the control system. In addition, such a layout also creates a human-centred manufacturing environment. The main benefits that can be expected from a cellular based manufacturing include simplified planning and control, reduced lead times, reduced WIP level, improved quality and increased job satisfaction.

However, there are a number of problems with this approach. One of these is that the proper functioning of the cells depend to a large extent to their initial definition. Therefore, at the detailed design stage cell configuration is frequently one of the most important tasks to be performed. The ideal cells should be completely autonomous, i.e., material flow between cells (known as the inter cell movement) is entirely eliminated. In real life this is not always achievable and the aims of cellular formation are usually to obtain maximum cell autonomy and in the same time minimize the number of bottleneck components (i.e., those which need to visit more than one cells) and bottleneck machines (i.e., those which are required in more than one cells – these can be eliminated through duplication, subcontracting or the use of a central service station). There are a large number of cell formation algorithms available, which can be classified into two broad categories, i.e., the product based methods and the product/process based methods. The former arrange components into sets according to their shape and size. The latter utilize production as well as product information to carry out cellular formation. Some of the product/process based algorithms are discussed in more detail in the following text.

Production Flow analysis

The Production Flow analysis [11, 12] was the first to utilize product/process information for cellular formation. It is an integrated approach which considers the group technology problem at a number of levels. At the factory and the group levels information of component flow and process are used to generate cell configuration. Material flow analysis utilizes operation and setup times to balance production, and finally, tooling analysis attempts to rationalise the tooling requirements.

Similarity coefficient method

The similarity coefficient based methods are a cell formation approach that has received significant attention recently. This is an approach which mainly operates on a component/machine matrix, as given by Eq.(1).

$$[a_{ij}] = \begin{bmatrix} a_{11} & a_{12} & \cdots & \cdots & a_{1n} \\ a_{21} & \cdots & \cdots & \cdots & a_{2n} \\ \vdots & & \ddots & & \vdots \\ a_{m1} & \cdots & \cdots & \cdots & a_{mn} \end{bmatrix}$$

(1)

where:

m = *the number of machines*
n = *the number of components*

$a_{ij} = \begin{bmatrix} 1 \ if \ components \ i \ needs \ machine \ j \\ 0 \ otherwise \end{bmatrix}$

$$s_{ij} = \frac{\sum_{k=1}^{n} \delta_1(a_{ik}, a_{jk})}{\sum_{k=1}^{n} \delta_2(a_{ik}, a_{jk})}$$

where:

s_{ij} = similarity coefficient between machs i,j

$\delta_1(a_{ik}, a_{jk}) = \begin{bmatrix} 1 \ if \ a_{ik} = a_{jk} = 1 \\ 0 \ otherwise \end{bmatrix}$

$\delta_2(a_{ik}, a_{jk}) = \begin{bmatrix} 0 \ if \ a_{ik} = a_{jk} = 1 \\ 1 \ otherwise \end{bmatrix}$

(2)

		Component/machine matrix (A)								$A*A^T$ matrix of machines							
		M/C No.									M/C No.						
		1	2	3	4	5	6	7			1	2	3	4	5	6	7
Cmp't	1	0	1	0	0	1	0	0	M/C	1	2	0	0	0	0	2	2
No.	2	0	0	1	1	0	0	0	No	2		1	0	0	1	0	0
	3	1	0	0	0	0	1	1		3			1	1	0	0	0
	4	1	0	0	0	0	1	1		4				1	0	0	0
	5	0	0	0	0	0	0	1		5					2	0	0
	6	0	0	0	0	1	0	0		6						2	2
										7							3

Figure 9 A NCC matrix

This method was first proposed by McAuley [13] with the so called single linkage similarity coefficient (SLINK) as defined by Eq.(2). The similarity coefficients are to be calculated for each The similarity coefficients are to be calculated for each of the machines (or machine groups), which can then be used by a cluster identification algorithm for cell formation. For instance, a threshold value of similarity coefficient can be used to identify the machines which should belong to a particular cell. One of the deficiencies of SLINK approach is that it fails to recognize the "chaining" problem caused by bottleneck machines – as with this analysis each machine can exist only in one group, there can be no duplication of machines [14, 15]. A number of attempts have been made to overcome the chaining problems including, for example, DeWitte's method [16], the average linkage similarity coefficient (ALINK) [17], and machine component cell formation (MACE) [18]. MACE, for example, is a simple extension to the similarity coefficients which uses three similarity coefficients to overcome some of the problems of single linkage cluster analysis. The method takes the component/machine matrix and produces a machine incidence matrix, which illustrates how many parts visit a pair of the machines. This is called the NCC (Number of Common Components) matrix, and is given by $A*A^T$ (see Figure 9).

$$ms_{i,j} = \frac{c_{ij}}{\min\left[\sum_{k=1}^{n} a_{ik}, \sum_{k=1}^{n} a_{jk}\right]}$$

where
c_{ij} = *number of common components using both machs*
$ms_{ij} = \begin{bmatrix} 0 \text{ if machines i and j are independent} \\ 1 \text{ if one is a subset of the other} \end{bmatrix}$

(3)

$$ps_{i,j} = \frac{c_{ij} \cdot c_{ij}}{\sum_{k=1}^{n} a_{ik} \cdot \sum_{k=1}^{n} a_{jk}} \tag{4}$$

$$fs_{i,j} = \frac{c_{ij} \cdot c_{ij}}{TF_i \cdot TF_j}$$

where TF_i = Total flow of components processed by mach i
(5)

From this matrix three coefficients are used to compute the similarities. These are respectively the additive similarity coefficient (Eq.3) based the total number of common components processed by a pair of machines, the product type coefficient (Eq.4) based on the total number of components processed by the pair of machines and the total flow coefficient (Eq.5) based on the total flow of the components. The first two coefficients are used initially to place the machines with the highest similarities alongside each other. This allows similar machines to be formed into groups. This data arrangement is then used and the inter-cell flows are calculated using the third coefficient, which determines the flow between the previously defined cells. From this the machines are arranged to minimise the inter-cell movement. The parts are then allocated to the cells by using a sorting routine.

Some of the later development have utilised a much wider range of manufacturing information including, for example, the sequence of operations and the expected work of the machines. In many cases, a_{ij}s are no longer restricted to the binary values of 0 and

1, but represent the duration of the operations in question. One example of these is provided by the production data based similarity coefficient proposed by Gupta and Seifoddini [19], as given by Eq.(6).

$$s_{i,j} = \frac{\sum\limits_{k=1}^{n} \left[\delta_{1,k} t_k^{ij} + \sum\limits_{o=1}^{n_k} \delta_{2,o} \right] m_k}{\sum\limits_{k=1}^{n} \left[\delta_{1,k} t_k^{ij} + \sum\limits_{o=1}^{n_k} \delta_{2,o} + \delta_{3,k} \right] m_k}$$

$$t_k^{ij} = \frac{\min\left(\sum\limits_{o=1}^{n_{ki}} t_{k,i}^{o}, \sum\limits_{o=1}^{n_{k,j}} t_{k,j}^{o} \right)}{\max\left(\sum\limits_{o=1}^{n_{ki}} t_{k,i}^{o}, \sum\limits_{o=1}^{n_{k,j}} t_{k,j}^{o} \right)}$$

where:

n = *number of components*
m_k = *planned prod. valume of part k during a period*
n_k = *number of times k visits both machs in row*
n_{ki} = *number of times k visits mach i*
$t_{k,i}^{o}$ = *optn time for k on machine i during oth visit*

$\delta_{1,k}$ = $\begin{bmatrix} 1 & if\ part\ k\ visits\ both\ of\ machines\ i,\ j \\ 0 & otherwise \end{bmatrix}$

$\delta_{2,k}$ = $\begin{bmatrix} 1 & if\ part\ i\ visits\ both\ machines\ in\ row \\ 0 & otherwise \end{bmatrix}$

$\delta_{3,k}$ = $\begin{bmatrix} 1 & if\ part\ k\ visits\ either\ of\ machines\ i,j \\ 0 & otherwise \end{bmatrix}$

(6)

As can be seen, this particular coefficient utilises the manufacturing information such as partwise production volume, pairwise production sequence and the operation time to define the similarity between a pair of machines. With these operational information taken into consideration, a more complete assessment of the situation can be expected. For instance, the inclusion of production sequence consideration in the formula recognises the fact that the material handling required to transfer a part between machines within a cell costs less than that needed to transfer the part between machines from different cells.

The ROC and its extensions

The rank order clustering (ROC) method [20] sorts the rows and columns of a component/machine matrix according to binary weighting. The weight of a row or a column are given by Eq.(7). The matrix is first sorted by the rows, and secondly by the columns. After the second step of this sorting cycle the initial order of the rows may be disturbed. The process is then repeated until there are no changes made to the orders of the rows and columns. An example of this iterative scheme is given in Figure 10.

$$Row\ i: \sum_{k=1}^{n} a_{ik}2^{n-k}$$

$$Column j: \sum_{k=1}^{m} a_{kj}2^{m-k}$$

(7)

a) Row ordering

	Component number 1 2 3 4 5	Row	Resultant
Binary value->	(16)(8)(4)(2)(1)	value	row order
1	0 0 0 1 1	3	
2	1 1 1 0 0	28	2–4–1–3
3	0 0 0 1 1	3	
4	0 1 1 0 0	12	

b) Column ordering

	Component number	Binary	Resultant
	1 2 3 4 5	value	column order
2	1 1 1 0 0	(8)	2–3–1–4–5
4	0 1 1 0 0	(4)	
1	0 0 0 1 1	(2)	
3	0 0 0 1 1	(1)	
Column value	8 12 12 3 3		

c) The resultant cells

	Component number 2 3 1 4 5	Resultant configuration
		Cell 1: components 1,2,3
2	1 1 1 0 0	machines 2,4
4	1 1 0 0 0	
1	0 0 0 1 1	Cell 2: components 4,5
3	0 0 0 1 1	machines 1,3

Figure 10 Rank order clustering

In comparison to other methods ROC is relatively simple and easy to implement. However, it has a number of limitations and does not guarantee to produce optimal solutions. ROC has been subsequently enhanced. The later versions include that of modified rand order clustering algorithm (MODROC) [21]. MODROC initially sorts the component/machine matrix using ROC, and then segments the matrix in the top left corner of the resultant matrix, as shown in Figure 11. The algorithm scans along the diagonal line to determine where the block which has been formed at this corner ends, and removes the identified block temporarily from the matrix. Next ROC is again applied on the reduced matrix and another block is sliced off. This process is repeated until no more slicing is possible. The pairwise similarities among the resultant blocks are evaluated by using similarity coefficients, so that some of these blocks may be rejoined. A similarity value of 1, for example, indicates that one of the block is a subset of the other (see Eq.3), and hence they can be joined together.

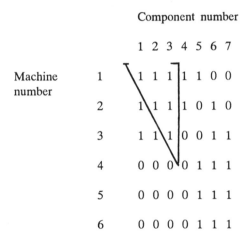

Figure 11 The slicing process of MODROC

The bond–energy algorithm

$$E = \frac{1}{2}\sum_{i=1}^{m}\sum_{j=1}^{n} a_{ij}(a_{i,j-1} + a_{i,j+1} + a_{i-1,j} + a_{i+1,j}) \qquad (8)$$

The bond–energy algorithm was first described by McCormic et al [22]. With this method the component/machine matrix is sorted using a measure of effectiveness, as defined by Eq.(8). The objective is to arrange the columns and rows of the matrix so that the maximum value of E is obtained, and in doing so the formation of a block diagonal (or the nearest thing to a block diagonal) is achieved. One of the weakness of this method is the long computation time that is often required. Improvements made to the original algorithm includes the work done by Bhat and Haupt [23], who used a matching matrix

$A*A^T$ (see Figure 9) as the basis for permutation rather than the original matrix A. Other attempts to utilise matrix $A*A^T$ include the Q analysis approach reported by Robinson and Duckstein [24].

Cluster identification algorithm

This was developed by Kusiak and Chow [25] which operates on the component/machine matrix as follows:

a. Set k=1 and generate the original matrix A_k.
b. Select any row i of A_k and draw a horizontal line.
c. For each entry of 1 on h_i draw a vertical line.
d. For each entry of 1 on v_j draw a horizontal line.
c. Repeat steps c and d till all crossing choices are exhausted.
d. All entries of 1 on a horizontal/vertical cross belong to machine cell k. Remove these entries from A_k to form A_{k+1}.
e. Set k=k+1 and repeat steps b–d, until A_{k+1}=0.

An example of this iterative method is given in Figure 12.

1) $k=1$, $i=1$, producing: h_1, and then v_4 and v_5, and then h_3; thus identifying entries of cell 1: $a_{1,4}$, $a_{1,5}$, $a_{3,4}$, and $a_{3,5}$.

		Component number 1 2 3 4 5	
Machine number	1	0–0–0–1–1–	h_1
	2	1 1 1 0 0	
	3	0–0–0–1–1–	h_3
	4	0 1 1 0 0	
		v_4 v_5	

2) $k=2$, $i=2$, producing: h_2, and then v_1, v_2 and v_3, and then h_4; no more iteration possible and thus identifying entries of cell 2: $a_{2,1}$, $a_{2,2}$, $a_{2,3}$, $a_{4,1}$, $a_{4,2}$, and $a_{4,3}$.

		Component number 1 2 3	
Machine number	2	1–1–1 ——	h_2
	4	0–1–1 ——	h_4
		v_1 v_2 v_3	

Figure 12. Cluster identification algorithm

This algorithm is simple and efficient in terms of programm coding and computational speed. To overcome the problem of bottleneck parts, an extended version of this method was later developed and reported by Kusiak [26].

Graph theoretic approach

This type of approach attempts to identify the weak relationships among the machines in terms of their similarity, and then rationalize the production routings to allow better groupings. First seeds (i.e., small subsets of the matrix) are defined which will act as the nucleus of cells to be defined. These seeds are then expanded by allocating their most similar rows and columns to them [27]. Like the similarity coefficient based approach, this method depends on a threshold value selected. The size of the cells must be defined by the user. One of such methods was developed by Vannelli and Kumar [28], which attempts to minimise the number of bottleneck components and bottleneck machines in the resultant configuration. With this method the components and machines are represented by a set of nodes, with their relationships shown by the appropriate linkages as defined by the component/machine matrix. The algorithm requires the user to define the maximum number of machines to be included in the cells, and then uses a simple formula to calculate the number of cells to be formed. The required number on seeds are then chosen on the condition that they must be disjointed. This can be achieved by using the $A*A^T$ matrix to find the entries which have no components in common. Once the independent seeds are identified, the rest of the entries is clustered around them using measures of effectiveness to analyze the quality of the resultant blocks. In addition, any sets of data which have connection with more than one seed is considered a bottleneck item and is removed from the set and put into a bottleneck set, so that appropriate action can be taken to produce completely disconnected cells.

ZODIC (Zero one data: ideal seed method for clustering) is another example of algorithms using this approach [29, 30]. ZODIC is based on the observation that for a manufacturing configuration with n components, m machines and v visits, there can be no more than k independent product families:

$$k \leq 1 + \frac{1}{2}\left[(m+n-1) - \sqrt{(m+n-1)^2 - 4(mn-v)}\right] \qquad (9)$$

The value of k, therefore, allows the initial number of cells to be determined. A number of k seeds can then be used to allow the entries of the matrix to be clustered around them. Three types of initial seeds can be used, i.e., the artificial seeds which represent the ideal configuration, the representative seeds which are subsets of the original matrix but identified according to an ideal configuration, and the natural seeds which are simply mutually dissimilar subsets selected from the matrix. Once chosen, the rows and columns of the matrix are assigned to these seeds by evaluating the "distance" of the rows (columns) from them (Eq.10).The resultant blocks may vary in size because some of the seeds attract less vectors than the others. Finally the optimal solution is obtained by diagonalizing the resultant blocks.

$$Dr_{ij} = \sum_{k=1}^{n} |a_{ik} - a_{jk}|$$

$$Dc_{ij} = \sum_{k=1}^{m} |a_{ki} - a_{kj}| \tag{10}$$

where:
   Dr_{ij} = *distance between row i and row j*
   Dc_{ij} = *distance between column i and column j*

Linear and integer programming approach

These methods attempt to construct multi–variable, linear or integer equations which define certain measures of effectiveness of a cellular based manufacturing system. Once a problem is modelled by a set of such equations, the optimal solutions can be sought for by using a computer based optimizing program, such as LINDO (Linear INteractive and Discrete Optimizer) [31]. For example, an integer programming model can be used to utilise the information given in a component/machine matrix to form a number of p cells [26]:

$$\min \sum_{i=1}^{n} \sum_{j=1}^{n} d_{ij} x_{ij} \tag{11}$$

subject to

$$\sum_{j=1}^{n} x_{ij} = 1 \qquad i = 1, 2, \ldots, n \tag{12}$$

$$\sum_{j=1}^{n} x_{jj} = p \qquad j = 1, 2, \ldots, n \tag{13}$$

$$x_{ij} \leq x_{jj} \qquad i = 1, \ldots, n, \; j = 1, \ldots, n \tag{14}$$

$$x_{ij} \leq 0, 1 \qquad i = 1, \ldots, n, \; j = 1, \ldots, n \tag{15}$$

where

   m = *numberofmachines*
   n = *numberofcomponents*
   p = *number of cells to be formed*
   d_{ij} = *distance measure between components i and j*

$$x_{ij} = \begin{cases} 1 & \text{\textit{if component i belongs to cell j}} \\ 0 & \text{\textit{otherwise}} \end{cases}$$

In this model d_{ij} is a measure of the distance between any two components. It can take a number of forms, including the one given in Eq.(10). The objective of this model is to minimize the total sum of such distances (also known as dissimilarities), under the constraints that each component can belongs to one cell only (Eq.12); exactly p cells are to be formed (Eq.13); and component i can belong to cell j only when this cell is formed (Eq.14). The end results from this particular model is rather like those of the seeding methods, in that a number of p nuclei components are identified and to which the rest of the components are attached to to form the end cells. For example, suppose 3 cells are to be generated from a set of 6 components. If components number 1, 3 and 4 are identified as the key components, then three cells C1, C3 and C4 will be formed accordingly, as shown below:

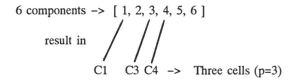

One advantages of this type of approach is that the objective function can be easily modified, and the constraints extended or relaxed so that the appropriate manufacturing information can be adopted to model a variety of situations. For instance, it is possible to construct a model in such a way so that the cost is minimized or utilization maximized [32, 33], or a component is allowed to visit a machine more than once [26]. However, the attractiveness of this type of methods decreases rapidly as the size of the problems involved increases.

The review given so far is by no means exhaustive – many other methods are available. Comparative study of the relative performance of these different approaches can be found in the literatures [34].

B. COMPUTER SIMULATION IN MANUFACTURING SYSTEMS ANALYSIS

Manufacturing computer simulation is a method for predicting the dynamic characteristics of an operation and thus improving the basis of a decision process. With a computer simulation model, a systems analyst or manager will be able to observe the behaviour of a process without the necessity of experimenting with the actual system. He may try out, for example, different planning policies, new operational conditions, new layout of equipment or different production capacities so as to evaluate the systems performance under various disturbances or to identify the bottlenecks. By using simulation the risks of introducing advanced manufacturing systems can be considerably reduced. As a result this methodology has been extensively used in the area of manufacturing systems studies both by academic researchers and practical engineers. It is very unlikely that anyone today would spend a substantial amount of money to build a new plant without having thoroughly checked its future performance by some sort of simulation study.

Table III Summary of simulation applications

USE OF MODEL	EXAMPLE OF APPLICATIONS	TIME HORIZON	TYPE OF MODEL	DATA NEEDED
strategic study at corporate level	market research and forecast, total capacity planning, financial evaluation, conceptual design and specification of manufacturing systems	long-term	both static and dynamic, mainly continuous but can be discrete, both stochastic and deterministic, always aggregated	highly aggregated at corporate level
production study at company level	inventory & capacity planning, master production scheduling, evaluation of production management decisions, conceptual design and specification of manufacturing facilities	medium- to long-terms	often dynamic, continuous or discrete, deterministic or stochastic at medium aggregation level	aggregate data at company level but more detailed than above, or reduced set of disaggregated data at shop level
production operations study at shop level	planning of detailed work loading, study of scheduling and job issuing policies, detailed planning and specification of manufacturing cells	short- to medium-terms	dynamic, discrete, deterministic models for on-line planning or stochastic for off-line research	highly detailed data at cell level

In relation to the design framework, two levels of simulation application can be distinguished, i.e., aggregated level at the conceptualizing stage and dis-aggregated level at the detailing stage. In the first, assessments of the system's gross operating characteristics are made, e.g., the overall means and variances of production rate and system capacities. For this type of information a relatively coarse model will suffice and a typical simulation run corresponds to a six to twelve months' production program. More detailed simulation models can then be used at the next level for the determination of equipment requirements, e.g., number of machines, number of transporters, number of pallets, machine configuration and layout, bottleneck problems and effects of equipment interference. The models used for these purposes need to handle more detailed system information and hence the simulation runs have to be restricted to a few days or a few weeks of actual production. Therefore, the choice of the modelling approach should consider the characteristics of the system in question and the nature of the problems to be tackled. In general the simulation models used in the area of AMS design can be distinguished between continuous and discrete, stochastic and deterministic, and aggregated and dis-aggregated. Their typical applications are summarized in Table III.

Among these models, discrete-event simulation is perhaps the most frequently used. Discrete-event simulation is concerned with the modelling of a system by a representation in which the state variables change at sudden distinct events – a typical phenomenon in a manufacturing operation. Although there is no general standard form of describing discrete-event modelling, the following terminology is often employed:-

The structural elements

    a.   **ENTITIES (or CUSTOMERS).** These are the basic temporary elements in the simulation model of a system. Jobs waiting in a shop to be processed are typical entities in a manufacturing system.

    b.   **ATTRIBUTES.** Each entity in the system may possess one or more attributes, which provide information about it. Job number, customer information, type of material, the colour and number of doors of a car body are some of the typical attributes encountered in a manufacturing environment.

    c.   **RESOURCES (or SERVERS).** These are the basic permanent elements of the model structure. They are responsible to provide services for the temporary elements. These typically include the machines and other work-stations in a manufacturing system.

    4)   **QUEUES.** A queue is formed in the system where entities wait for the service of a resource. An in-process buffer store gives on example of queues.

The dynamic elements

    5)   **EVENTS.** These are the points in time at which changes of system state occurs, such as when a machine is loaded and becomes busy or when it finishes a machining operation and becomes free.

    6)   **ACTIVITIES.** The operations initiated at each event are called activities. Activities are responsible for the transformation of the states of system entities.

7)   **PROCESSES.** A process is simply a group of sequential events. For instance, a raw material arrives at a workstation, is machined or processed, and then goes to the next workstation. In general, a process describes a part or complete experience of an entity as it flows through the system.

8)   **SIMULATION CLOCK.** This is a variable in a simulation model used to keep track of the current value of simulated time as the simulation proceeds.

The general form of a discrete event model may be outlined according to the above definitions. It can be assumed in general that a system would consists of I **RESOURCES** S_i (i = 1, 2, ..., I), with each of them preceded by a **QUEUE** Q_i. These **RESOURCES** and **QUEUES** could be interconnected in an arbitrary way so that a particular system under study can be presented. During the dynamic simulation process, the temporary **ENTITIES** E_j (j = 1, 2, ..., N) arrive at **RESOURCES**; after one or more **ACTIVITIES** taking place, complete their relevant transformation there, would normally proceed to some next **QUEUE** Q_k, and hence to **RESOURCE** S_k. According to the state of Q_k (empty or non–empty), the status of S_k, and possibly the priority attribute of E_j, a waiting delay $w_{j,k}$ would in general be experienced by E_j at this stage ($w_{j,k}$ may have the value of zero). When it is E_j's turn to be served by S_k, a processing delay $p_{j,k}$, which is one of the primary system parameters, would again be experienced by E_j. When E_j leaves S_k, and if S_k still remains functional, the next **ENTITY** may then proceed to S_k and the process continues. If there is no next **ENTITY**, **RESOURCE** S_k stays idle until the next **ENTITY** arrives. Additional delays may occur when some of the **RESOURCES** occasionally become unfunctional. This dynamic process would continue until a certain predetermined termination condition is met, e.g., when the simulation clock has reached the required point of simulated time; when the whole set of **N ENTITIES** has passed through the system; or when some other criteria have been satisfied. The operation of a machining station is a good example. When a workpiece arrives for machining, it will wait at a buffer store if the required machine is not free. Once the machine is ready to take a job and there is a job waiting in the buffer, the machining process can start. Therefore, the state of the system changes discretely from one state to another through time. From a system analyst's point of view, the only concern with the machine is whether the machine is busy, idle, or broken down, rather than how the machine actually processes the job. The time taken for the machine to process a workpiece can either be sampled from some appropriate distribution (random simulation), or set to a known constant (deterministic simulation). This is also true for other activities in the system. Therefore, the next change in state of the system can be simulated by referring to these known activity times. It can be seen therefore that the general problem of discrete event simulation modelling is how to embody all of the interactions of the above mentioned structural and dynamic elements.

Almost all simulation languages use one of a few basic approaches to accomplish this process, with each embodying a distinctive view point. However, these simulation languages inevitably provide some common features such as model building and editing facilities, graphic animation and interactive control, and results analysis and report generation facilities. These have made the programming tasks of a simulation study relatively easy, because special knowledge about computer programming of simulation logic is no longer a prerequisite so that this type of computer–aided system analysis can be carried out by those analyst who are not computer programming experts. Another attractive features of the modern simulation language is the facilities of graphics animation. Systems layout, elements and movement of entities are represented either by icons or characters in an animation simulation model. Most of the animation system restrict the movement of entities in a two–dimensional plane. However, there are a number of systems which allow entities to move within a three–dimensional space.

The simulation languages used for manufacturing systems analysis may be classified into two categories, general packages and special packages. The general application system can be used for simulating processes in various areas. They only provide basic simulation functions and statistic facilities. Examples of this class include GPSS (General Purpose Simulation System) developed by IBM, HOCUS (Hand Or Computer Universal Simulator) developed by P–E Incubucon, SIMAN/CINEMA developed by Systems Modelling Co. and PCModel (PC Modelling system) developed by SimSoft Ltd. The special systems for manufacturing simulation are specifically intended for modelling manufacturing situation, and therefore reduce the programming time and effort. Examples of this type of simulation systems include:–

MAST (MAnufacturing System design Tool) developed by CMS Research Inc. – this system is particularly suitable for the simulation of computerized manufacturing systems

SIMFACTORY which is used specifically for simulating for Job–shop systems

WITNESS II which may be used for both Job–shop and computerized manufacturing systems simulation

Information about the currently available computer simulation systems, including the name of packages, detail of their supplier, application summaries, hardware and systems requirements as well as the current price can be found in the appropriate professional journals. Figure 13 gives two examples of simulation model for a simple machining operation which are written respectively in GPSS and SIMAN – two of the well–established simulation languages available.

As can be seen from the previously discussion, advanced manufacturing systems involve a total systems concept which embraces a wide range of manufacturing aspects from customers enquiry to product delivery, involving all of the operational functions. In order to develop the strategies and policies regarding the efficient operation of AMS, it is no longer sufficient to study only artificially de–coupled hypothetical and low dimensional systems. This will require an analysis approach which is more realistic and problem–specific. Therefore a relatively large–scale computer model capable of simulating the operation of a proposed manufacturing in its systems context is likely to be a valuable decision aid. In the past, large complex models have often been criticized for requiring large amount of both manpower and computing resources to develop, making them too expensive and time–consuming to be of any real use. This kind of criticisms are often well founded. However, when it comes to the design and evaluation of AMS, the use of such simulation models would seem to be readily justifiable due to the need to truthfully reflect the complex interactions inherent in a AMS environment, and the long–term and expensive investments involved.

==
A simulation model of a machining operation written in GPSS.

```
GENERATE  300,100          ;create next part
QUEUE     BUFFER           ;part waits in queue
SEIZE     MACHINE          ;attempt to grab the m/c
DEPART    BUFFER           ;engage m/c and end of waiting
ADVANCE   400,200          ;the operation takes certain time
RELEASE   MACHINE          ;operation done, free m/c
SAVEVALUE AVE_Q,QT@MACHINE ;collect queue data
TERMINATE 1               ;part leaves system
```

A simulation model of the same machining operation written in SIMAN.

```
Model file:-
BEGIN;
  CREATE:RN(1,1):MARK(2);    create next part
  QUEUE,1;                   wait for processing
  SEIZE:MACHINE;             attempt to grab m/c
  DELAY:UN(3,1);             engage m/c for certain time
  RELEASE:MACHINE;           operation done, free m/c
  TALLY:1,INT(2);            collect resulting time
END;
```

```
Experiment frame:-
BEGIN;
  PROJECT,MACHINE,MCH,28/2/1990;         title
  DISCRETE,80,3,1;                       system organisation
  RESOURCES:1,MACHINE,1;                 define capacity of machine
  TALLIES:1,PART STAY;                   define time to be collected
  PARAMETERS:1,300,100:2,0:3,400,200;    distribution   parameters
DSTAT:1,NQ(1),Q_LENGTH:2,NR(1),M_UTIL;   output data
  REPLICATE,1,0,20000;                   simulation run time
END;
```
==

Figure 13. Example simulation models

C. EXPERT SYSTEMS AND SYSTEMS DESIGN

An expert system is basically a software program that contains knowledge and processing rules with the purpose of emulating the decision–making process of an expert in a particular area. In many of its applications areas, an expert systems is very similar to a computer simulation in that it is also a means of making decisions in complex situations in which mathematical techniques are difficult or impractical to apply. However, expert systems usually give **prescription**, whereas simulation models provide **prediction**. That is, unlike a simulation model, an expert system suggests a course of action and provides a rationale or explanation for the suggested course of action. Consequently, for certain applications the traditional simulation or analytical tools alone do not compare with the benefits of knowledge based expert systems in terms of prescription and explanation. On the other hand, expert systems are not constructed to deal with the complex dynamic behaviour of large systems.

Recently there are increasing interest in expert systems for solving manufacturing problems. So far as systems design is concerned, the following categories of application can be identified:

> Knowledge–based simulation,
> Knowledge–based product design,
> Knowledge–based process planning,
> Knowledge–based machine specification,
> Knowledge–based equipment selection,
> Knowledge–based facility configuration, and
> Knowledge–based facility layout.

There are two basic ways in which expert systems can be applied, i.e., the stand–alone approach and the tandem approach [26]. A stand–alone system contains a knowledge base consisting of sufficient information and rules about a specific problem domain so that a problem can be solved through a reasoning process without the need for additional analytical effort. The knowledge contained in such a system can usually be classed into domain knowledge which is related to the facts and rules, and problem–solving knowledge which is used as an interpreter and a scheduler. The design task of machine specification, for example, is an typical problem which has been tackled through this approach. Based on a set of initial requirements in terms of capacity, tolerance and cost, etc., the suitable machine types currently available on the market may be located using a purpose–built expert system on its own. The tandem approach, on the other hand, requires the concurrent use of both expert system and conventional modelling approach for problem–solving. Knowledge–based simulation models are typical examples of this tactic. There are two main applications of expert systems in simulation modelling, i.e., to help the user to build a simulation model and carry out experimentation, and to interpret the results produced so as to make further refinement.

It is believed that the future uses of expert simulation systems within manufacturing will be extensive, for they have the potential to allow manufacturing systems to be evaluated and analyzed under various design changes, alternative control actions, different procedures and new policies with the minimal requirement on expert intervention. An ideal knowledge based simulation system should be able to:-

* Accept a problem description and synthesize a simulation model by consulting an appropriate knowledge base,

* Evaluate alternative scenarios and choose the one that best satisfies the objectives, and

* Explain the rationale behind any decisions made and learn from past experience.

For simulation and expert systems to be integrated in the first regard, in addition to the traditional need for a simulation model to logically express both the static structure and the dynamic behaviour of a system being modelled, there is a also a need to link the model structure into an environment which is suitable for an knowledge–based program generator to work in. Consequently, for the automatic construction of simulation models two main bases are required: a model base which contains the executable numerical sub–models and a knowledge base that contains the semantic information necessary for the use of these models. One way to realise this, for example, is to build a simulation model by selecting and then arranging system entities and procedures from a standard library through an expert system which interprets user requirements. In relation to this approach, one scheme for the knowledge representation of simulation modelling is known as object oriented representation in which every element of a model may be thought of as an object that is described by its properties [35]. These properties may be descriptive, behavioral, structural or taxonomical (examples of object orientated languages in use include C++, SMALLTALK and LOOP). The application of expert systems for the automatic modelling in simulation is referred to as the intelligent front end (IFEs). There are basically three alternative approaches to automate the traditional information acquisition and model programming tasks, as shown below:

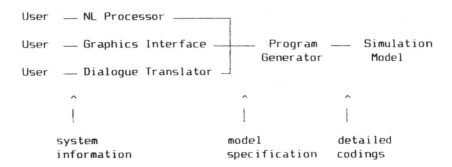

As shown, a detailed definition of a simulation model may be obtained from a natural language description of the system to be simulated by a user through a natural language processor. This process may also be carried out through a graphics interface, which allows the structure of the layout of components to be described. Similarly this can also be achieved through a dialogue translator, which obtains the necessary information through question/answer communication. Once the detailed information has been obtained by the system through these processes, the model itself can then be generated by an automatic programming system. Each of these approaches interactively produces the detailed model specification to satisfy the users' requirements.

In the second application area, the value of expert systems becomes clear when it is realised that while simulation is normally used to evaluate suggestions, expert systems on the other hand make suggestions based on evaluation. Operating together, therefore, skilled manual intervention can be reduced and a solution arrived at more reliably and more rapidly. For instance, one of the major disadvantages identified of the simulation approach is that only one set of circumstances can be tested by any individual simulation run. An expert system designed and integrated into the modelling frame could overcome this by selecting which variables are likely to give a desired effect, and how they should be altered and controlled. Such a system could then adjust the variables and repeat the simulation, iterating until an optimum is found. An expert system used for this particular purpose is referred to as an intelligent back end (IBE).

One of the major disadvantage of using expert systems in a manufacturing environment is due to the diversity of manufacture. In most of the cases when an expert system is to be used in simulation, it must first be specially developed and tested before incorporation, hence adding significantly to the time and cost of the project. It may be however that there exist a set of problems and situations which have broadly similar properties, and while requiring different simulation models may be able to use a common expert system to suggest qualitative improvements. Fortunately some of the common problems involved in manufacturing systems design, such as cell configuration, line balancing and equipment layout, do seem to belong to this category and hence a general design framework together with an integrated information system can be a feasible proposition. For example, although it is possible to obtained optimal solutions with some of the previously listed algorithms, there are always some kind of difficulties or pitfalls to apply them in practice. For instance, the 0/1 matrix based algorithms for cellular formation are easy to understand and simple to use. However, they must be applied with caution, because a number of key production parameters, such as machine capacities and operation times are ignored. On the other hand, those algorithms which do attempt to take these factors into account tend to be very complex, and hence extremely difficult to adopt in practice due to the amount of information needed. One way to overcome this problem is to use these algorithms interactively with the design framework outlined in Section II together with a CAMSD database linked to simulation and expert systems. A possible scheme for this integration will have the following four distinct entities:

a)   *Algorithm base.* This is a computer program which, once the appropriate information are provided, is capable of carrying out the analytical process of a selection of algorithms.

b)   *Data extraction.* This requires a program which is capable of extracting the necessary information from the database and present them in a suitable format for the particular algorithms to be used. Conventionally this kind of task will have to be accomplished through custom written routines. It is possible, however, to use standard (or widely accepted) file format to gain access to a wide range of systems.

c)   *Data estimation through simulation*. The functionally most suitable algorithms may require manufacturing information which are not yet available, making their application in real life impossible. However, as simulation modelling is to be used as an integrated element of the framework, a plausible approach is to use this to estimate the necessary data, and hence make the application of these multi–variable based algorithms practical.

d)   *Knowledge based execution*. By utilizing a knowledge base of cellular formation, an expert system will be able to coordinate the functions involved in (a) – (c) above. For a particular situation, this ensures that the suitable models and data are applied, and necessary iterative processes are initiated so as to guarantee the optimal solution.

Such a scheme can also be applied to the other design tasks with the framework.

VI. CONCLUSION

This completes the whole picture of the CAMSD framework. The type of products to be manufactured, whether they are pulled by demand or pushed by new technology, has been used as the basis of a framework of manufacturing systems design. In relation to this framework, the structure of a CAMSD database, which allows a designer to input and effectively use the data about the current manufacturing system (if there is one) and the details about the required manufacturing system, is also discussed. These, together with a discussion on the currently available computer–aided tools, are aimed to improve the quality of systems design and specification. If it is first properly designed and assessed, the implementation and operation of the actual manufacturing system concerned should be relatively trouble free and provide good returns on the capital invested.

References

1.   B. Wu, "Fundamentals of Manufacturing Systems Design and Analysis," Chapman and Hall, London, 1991

2.   F. Choobineh, "A Framework for the Design of Cellular Manufacturing Systems," International Journal of Production Research, Vol.26, Taylor & Francis Ltd, 1989

3.   N.D.C. Slack, "Achieving a Manufacturing Advantage," W.H. Allen, 1990

4.   W. Skinner, "The Focused Factory," Harvard Business Review, May/June 1974

5.   E. Downs, et al, "Structured Systems Analysis and Design Method – Application and Context," Prentice–Hall Inc., 1988

6.   N. Ajderian, "Do You Really Want That New Machine?" Professional Engineering, November 1989

7.   K. Swann and W.D. O'Keefe, "Advanced Manufacturing Technology: investment decision process – part 1," Management Decision, No.1, 1990

8.   K. Swann and W.D. O'Keefe, "Advanced Manufacturing Technology: investment decision process – part 2," Management Decision, No.3, 1990

9.   R. Wild, "Production and Operations Management," Cassell Education Ltd., 4th Ed., 1989

10.   R. Elmasri and S.B. Navathe, "Fundamentals of Database Systems," The Benjamin/Cummings, Inc., California, 1989

11.   J. Burgidge, "Production Flow Analysis," Production Engineer, Dec. 1963

12.   J. Burgidge, "A Manual Method for Production Flow Analysis," Production Engineer, Dec. 1979

13.   J. McAuley, "Machine Grouping for Efficient Production," Production Engineer, Feb. 1972

14.   J.R. King and V. Nakornchar, "Machine Component Group Formation in Group Technology – Review and Extension," International Journal of Production Research, Vol 20, 1982

15.   H. Seiforddini, "A Note on the Similarity Coefficient Method and the Problem of Improper Machine Assignment in Group Technology Applications," International Journal of Production Research, Vol 26, 1989

16.   J. De Witte, "The Use of Similarity Coefficients in Production Flow Analysis," International Journal of Production Research, Vol 18, 1980

17.   H. Seifoddini, and P.M. wolfe, "Application of the Similarity Coefficient method in Group Technology," IIE Transactions, Vol. 18, 1986

18.   P. Waghodekar and S. Sahu, "Machine Component Cell Formation in Group Technology: MACE," International Journal of Production Research. Vol 22, 1984

19.   T. Gupta and H. Seiforddini, "Production Data Based Similarity Coefficient for Machine–Component Grouping Decisions in the Design of a Cellular Manufacturing System," International Journal of Production Research, Vol 27, 1990

20.   J.R. King, "Machine–Component Group Formation in Production Flow Analysis: An Approach Using a Rank Order Clustering Algorithm," International Journal

of Production Research, Vol 18, 1980

21.   M.P. Chardrasekharan and R. Rajagopalan, "MODROC: an Extension of Rank Order Clustering for Group Technology," International Journal of Production Research, Vol 24, 1986

22.   W.T. McCormick, P.J. Schweitzer, and T.W. White, "Problem Decomposition and Data Reorganisation by a Clustering Technique," Operations Research, Vol 20, 1972

23.   M.V. Bhat and A. Haupt, "An Efficient Clustering Algorithm," IEEE Transactions on Systems, Man, and Cybernetics, Vol SMC–6, 1976

24.   D. Robinson and L. Duchstein, "Polyhedral Dynamics as a Tool for Machine Part Formation," International Journal of Production Research, Vol 24, 1986

25.   A. Kusiak and W.S. Chow, "An algorithm for Cluster Identification," IEEE Transactions on Systems, Man, and Cybernetics, Vol SMC–17, 1987

26.   A. Kusiak, "Intelligent Manufacturing Systems," Prentice–Hall, Inc., New Jersey, 1990

27.   R. Rajagopolan and J.L. Batra, "Design of Cellular Production Systems – A graph Theoretical Approach," International Journal of Production Research, Vol 13, 1975

28.   A. Vanelli and K.R. Kumar, "A Method for Finding Minimal Bottleneck Cells for Grouping Part Machine Families," International Journal of Production Research, Vol 24, 1986

29.   M.P. Chandrasekharan and R. Rajagopalan, "An Ideal Seed Non–Hierarchical Clustering Algorithm for Cellular Manufacturing," International Journal of Production Research, Vol 24, 1986

30.   M.P. Chandrasekharan and R. Rajagopalan, "ZODIAC – An Algorithm for Concurrent Formation of Part–Families and machine cells," International Journal of Production Research, Vol 25, 1987

31.   L. Schrage, "Linear Programming Models with LINDO," The Scientific Press, California, 1980

32.   F. Choobineh, "A Framework for the Design of Cellular Manufacturing Systems," International Journal of Production Research, Vol 25, 1988

33.   H.C. Co and A. Arrar, "Configuring Cellular Manufacturing Systems," International Journal of Production Research, Vol 26, 1988

34.   S.M. Shafer and J.R. Meredith, "A Comparison of Selected Manufacturing Cell Formation Techniques," International Journal of Production Research, Vol 28, 1989

35.   R.E. Shannon (1989) "Knowledge-based simulation techniques for manufacturing" in "Knowledge-Based Systems in Manufacturing," (A. Kusiak, ed.), Taylor & Francis, 1989

SOFTWARE CONFIGURATION TECHNIQUES IN OPERATIONAL SYSTEMS

M. Sloman, J. Kramer, J. Magee, P. Butryn†

Department of Computing,
Imperial College of Science Technology and Medicine,
180 Queen's Gate, London SW7 2BZ.

† BP Research, BP International Ltd., Sunbury Research Centre,
Chertsey Road, Sunbury on Thames, Middx. TW16 7LN

I. INTRODUCTION

There has been considerable research into programming languages which emphasise modularity and the reuse of components, for example ADA, SR, Argus, NIL, Emerald. However, there has been comparatively little work on languages which emphasise the separation of programming individual components *(programming in the small)* from the specification and construction of distributed systems from predefined, reusable components. *(programming in the large)* [deRemer 76]. The various languages mentioned above do have some support for combining separately compiled modules, but they treat the resultant system as a single large program. The specification of a current system is buried within the current state of a program which makes it difficult to modify or extend the system.

The diversity of the software components useful in large distributed systems implies that appropriate, state-of-the-art programming languages should be used. For instance, high-level procedural languages (eg. C, Pascal, Modula2) should be used for the real-time control and monitoring, object-oriented languages (eg. Smalltalk, C++) for man-machine interfaces, artificial intelligence languages (eg. Prolog, Lisp) for the expert systems and knowledge bases needed for the advanced intelligent (ie. decision making) components, and Fortran for the existing numerical analysis packages.

CONTROL AND DYNAMIC SYSTEMS, VOL. 47

Although the rewards are great, the use of heterogeneous programming languages exacerbates the problems of integration, as they may make use of incompatible communication mechanisms and data representations.

This paper describes a configuration approach to integrating heterogeneous software components which communicate and interact in a distributed computer automation system. The Conic configuration facilities support both building of an initial system and the subsequent dynamic modification of the running system to permit reconfiguration for operational changes as well as evolutionary changes to incorporate new functionality. This configuration approach involves specifying a system in terms of the instances of components; the interconnection between component interfaces; the mapping of software to hardware and modifying systems. The emphasis is more on the structural relationships between components and this provides an abstraction in which changes to a system in terms of adding or removing component types, instances and interconnections can be easily understood and formulated [Goguen 86]. Support for dynamic reconfiguration is particularly important for operational systems which cannot be shut down in order to make changes.

Motivation

Operational Automation systems may involve many different microcomputers running a variety of distributed programs. Automated tools are needed to cope with the complexity of constructing the underlying support services as well as individual applications.

A **configuration specification** of a system in terms of components and their interconnection provides a high level description of the logical structure of a distributed system, from which interactions and dependencies can be determined. Such a specification can be used by automated tools for constructing a system from the set of components.

A distributed system is likely to have multiple instances of a particular software component type located at different computers in the system. In order to permit this reuse of software components in different configurations, all configuration dependent information such as binding between component instances and location of components must be determined at configuration or run-time. Using a configuration language to specify interconnection information (i.e. the bindings between component interfaces) results in a configuration specification which holds information on component dependencies and so it is very easy to determine what components will be affected by a change. In addition validation of type or behaviour compatibility can be performed at configuration time rather than run-time. In control and automation applications, processors may have dedicated functions or particular devices connected which determine where components must be located. This means the system designer must provide location information for software components. Also performance considerations may determine whether components need to be co-located.

A configuration language provides a suitable level for specifying location constraints for reusable components.

The configuration level abstracts from the algorithmic details and implementation issues which are part of the component programming level. This makes it easier to provide tools which perform validation and analysis at this level. For example including timing constraints within the configuration specification permit a static analysis of a distributed applications real-time behaviour [Coulas 87 Barbacci 88]. This appears to be the most appropriate level for both specifying and validating behaviour, but very little work has been reported on this.

Automation systems can have a lifespan of many years during which they do not remain static. **Operational changes** can be predetermined and include the reconfiguration required to cope with time-varying usage patterns or reconfiguration after failures. In addition the system must evolve to incorporate new functionality, in response to changing requirements, or to incorporate new technology. These **evolutionary changes** require the ability to install new types of components, and remove old ones. Although operational changes could be dealt with at the component programming level, the evolutionary changes cannot be predetermined so must be part of the configuration programming level. Both evolutionary and operational change is part of the management of a distributed system. Managing a system requires access to a clear specification of what is being managed and a flexible means of specifying changes to the system to cope with the above evolutionary and operational changes. It may not be feasible to shut down an application in order to change it so support tools are needed both for building initial systems and subsequently **dynamically reconfiguring** them.

The above discussion has shown that a clear and precise **configuration specification** of a distributed application is needed in terms of the components within it, the interactions between components, interactions with the environment and the mapping of software components onto hardware. A similar specification of the underlying support environment (operating system) is also needed. A language based approach to configuration specification which is clearly separated from the component programming languages meets these requirements [Kramer 85, LeBlanc 85, Schwan 86]. In this paper we identify the requirements and characteristics of configuration languages and discuss the issues relating to dynamic configuration. The Conic Configuration tools will be used as an example.

Section II of the paper discusses requirements for configuration from the viewpoints of both component programming and configuration languages. Sections III and IV describe the facilities supported in Conic for static and dynamic configuration. Approaches to integration of heterogeneous components are described in section V, followed by a case study of the use of these tools in a petrochemical plant. In section VI, we discuss the current work on a new configuration language being developed and conclusions of our experience with the use of Conic is given in section VII.

II. CONFIGURATION LANGUAGES

II.A Component Programming Level Prerequisites

There are a number of properties which have to be satisfied at the component programming level in order to permit the separation from the configuration programming level [Kramer 85].

Modularity the language must support the programming and compilation of software components which are independent of the configuration in which they are used. This *context independence* can only be achieved if the the statements inside a component name local entities. If a component directly names another component then it cannot be reused in a different configuration without changing the name in the component and recompiling it.

Component types - components should be treated as types (templates) from which multiple instances can be created. It permits reuse of the component specifications. For example there will be a need for many sensor device handlers in a monitoring system. It is useful to be able to pass instantiation parameters to an instance to tailor it for the particular context in which it is being used e.g. to pass a device address to a device handler.

Strict Interfaces - the only interaction with a component should be via a clearly defined interface. This provides information hiding and shields users of the component from changes to its implementation which do not affect the interface. In order to fully define the behaviour of a component, the interface must specifiy both operations *provided by* the component and the operations it *requires* from other components (c.f. input and output signals for an integrated circuit).

Communication transparency - the component interaction primitives should have the same syntax and semantics for both local and remote communication. There should be no differences, other than performance for interactions between components in the same computer or in different computers connected by one or more networks. This permits a logical configuration specification to be mapped to different hardware configurations.

Interface Binding - components can only interact via their interfaces. There must be a binding or association between interfaces on component instances to permit interaction between them. In order to support configuration independence this binding must be at configuration time or run time and not at compile time. Most remote procedure call implementations provide an operation to bind an imported interface to an exported one [Birrell 84]. This run time binding is normally instigated by a client (importer) binding to a server (exporter) but there is also the need for transparent rebinding instigated by the system rather than the client for example to recover from a failure.

The proponents of object oriented languages claim that they meet some of the above requirements [Wegner 90]. However inheritance, which is a key feature of these languages, can result in hidden interfaces between types in the type hierarchy [Snyder 86]. The operations which can be invoked on external objects are not explicitly defined as part of the interface with many object oriented languages. Objects are created dynamically within the programming language and bindings can occur by assigning object references, received as parameters within messages, to object variables within a program. Thus there is no clear separation between the programming and configuration level.

II.B Configuration Language Properties

A clear and manageable Configuration Specification is needed by programmers defining application or system software, by management which is responsible for maintaining and evolving a system and by the underlying support tools which are used to construct the initial system and perform subsequent changes to it. The required properties of a language for specifying configuration are discussed below.

A configuration language should be *declarative*. If configuration statements are embedded within a procedural program, the current state of the configuration depends on the current state of the program and that is not easily determined. A declarative specification is more amenable to analysis and validation with decisions about the order in which operations are performed being left to the system. The underlying support system can then exploit the inherent parallelism of creating components on multiple distributed computers. Start up ordering of components can be achieved at the component programming level using synchronisation primitives to interact with the configuration management system (see section IV.B).

In order to provide the flexibility of constructing a variety of systems from the same reusable component type, the configuration should be specified in terms of *system structure* - creating instances and binding of interfaces [Goguen 86]. This is analogous to building different hardware systems from the same basic integrated circuit components.

Hierarchical Composition - it should be possible to encapsulate multiple components within a single **composite component** [Magee 87, LeBlanc85]. This form of abstraction is essential to cope with the complexity of large systems. The implementation of a support environment or particular service can then be encapsulated and hidden behind a set of interfaces. This composite component should have the same characteristics as a single component i.e. it should be a template from which instances can be created and have a clearly defined interface. A composite component template defines the templates required, component instance created from those templates, plus instantiation parameter values passed to the instances.

Interface Binding - as mentioned in II.A, there is a need for third party, configuration level binding to implement failure transparency [Friedberg 87]. Similarly there is a need for configuration time binding of a component to its support environment. This is equivalent to binding to instances which are predeclared and is likely to be a static binding for the lifetime of the component, unless component migration is supported. Another aspect to interface binding is that it should be possible to perform compatibility validation - at least type checking, but behavioural compatibility validation is also desirable but difficult to do.

A configuration language needs access to interface specifications. The two main approaches to *Interface Specification Languages* (ISL) are based on procedure names plus call and return parameters [Jones 85], or ports plus messages types which will be described in section III.A. Although the two approaches differ in syntax, they essentially provide the same facilities - local nameholders to represent a remote entity (port or procedure names) and a specification of the information transferred (parameters or message types). Message primitives may be synchronous (request-reply) or asynchronous (unidirectional) and some of the recent procedure based ISLs also support asynchronous procedure calls as well as the traditional synchronous calls [Ansa 89].

There are two paradigms for configuration specification languages:

i)   A graphical language which is based on icons, menus and pointing devices emphasises the spatial relationships between components and provides an ideal human interface for specifying configurations.

ii) A textual language is the traditional approach and is more amenable to machine manipulation. It is better for expressing very complicated interconnection patterns (e.g. based on a function call to a random number generator) and repetitive instances such as arrays of identical components.

We believe it is necessary to be able to easily transform to and from textual and graphical specifications to enable the most appropriate paradigm to be used for different aspects of the configuration specification. Examples of these paradigms within the Conic environment are given in section III.

Configuration programming is needed at various levels - to specify a composite component template which encapsulates instances of other component types; to specify the support system (operating and communication system) from instances of components; to specify the components which form an application and to specify the binding of an application to its environment. We have explained the concepts and requirements for configuration programming and now describe an example of an environment which supports these facilities.

III CONFIGURATION PROGRAMMING IN CONIC.

The Conic environment [Kramer 85, Magee 87], developed by the Distributed Software Engineering Group at Imperial College, provides support for configuration programming for distributed and concurrent programs. The environment provides support for two languages, one for programming individual task modules (processes) with well defined interfaces, and one for the configuration of programs from groups of task modules. (Conic uses the term *module* for a component). Conic supports both textual and graphical configuration languages. In addition the environment provides support for the reuse of components in different contexts and support for dynamic configuration. This latter facility is achieved using on-line management tools which permit dynamic creation, control and modification of application programs. The basic Conic environment has been in use for over 8 years. It has amply demonstrated the utility of configuration programming.

III.A An Example

Fig. 1 shows a simple water supply system which will be used as an example to illustrate the features of the Conic Configuration Environment. A user consumes water from a reservoir which is fed from a lake by a pump. The pump switches on when the water level in the reservoir reaches a low level detector and switches off when the water level is high. Fig. 2 shows the Conic modules required to simulate this system.

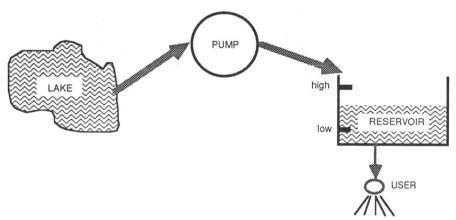

Fig. 1 A Simple Water Supply System

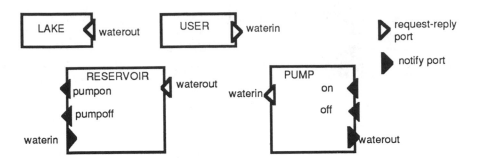

Fig. 2 Water System Task Modules

A request-reply port represents a bidirectional synchronous message transactions and a notify port represent a unidirectional asynchronous transaction.

III.B Task Module Types

Task Module types are defined using the Conic programming language which is based on Pascal with extensions for modularity and message passing. These are the most primitive components with respect to configuration and are sequential "lightweight" processes. The interface to a module is defined by typed exitports and entryports. Messages are sent out of a module via exitports and received into modules from entryports. Messages can be of any Pascal datatype (except pointer types for obvious reasons). The type definitions are imported from definition units by the **use** clause. Fig. 3 gives an outline of the Conic program for the task module reservoir. This module either receives a message with water from the pump or if there is sufficient water in the reservoir it will accept requests from a user to take water out. The module is parameterised with its capacity.

```
task module reservoir (capacity:integer);

    use pumptypes: volume
    exitport                          {INTERFACE}
        pumpon,
        pumpoff : signaltype;
    entryport
        waterout : volume reply volume;
        waterin : volume;
    var   reqvolume, level : volume;      {LOCAL DATA}
        ....


    begin                             {TASK BODY}
        initialise;
        loop
            select
                receive water from waterin
                    => { adjust  water level & stop pump if high!}
            or
                when (level>0) receive reqvolume  from waterout
                =>
                    if reqvolume > level
                    then reply  level to waterout
                    else reply  reqvolume to waterout
                    {reduce level, start pump if water level low}
            end;
        end;
end.
```

Fig. 3 Reservoir Task Module

III.C Conic Configuration Language

We can construct an initial water supply system consisting of one instance of each of the module types shown in Fig. 2 and interconnecting their exit- and entryports. The links between exitports and entryports is an interface binding to allow modules to communicate by message passing. The Conic environment permits only ports of the same type to be connected. Multiple exitports can be linked to a single entryport which is particularly useful for binding clients to servers. A single exitport can be linked to multiple entryports to allow multi-destination message transfer. The configuration program for this initial system is shown both textually and graphically in Fig. 4.

The system is actually created by submitting the configuration description to a configuration manager tool which downloads module code into target processors or instantiates processes under Unix as appropriate. The configuration management tool and its supporting environment is described in [Magee 87]. The configuration description may be submitted either as text or using the graphical interface described in more detail in [Kramer 89]. As can be seen from Fig. 4, the graphical description contains less information than the textual one. Additional information is elicited by the graphical interface through the use of dialogue boxes. This extra information consists of module location (the **at** clause) and module parameters.

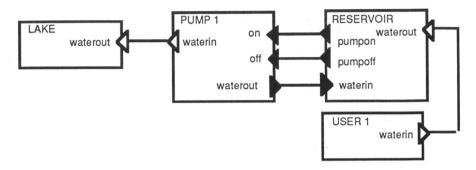

a) Graphical Representation

system watersystem 1;

use   lake, pump, reservoir, user;   {type templates}

create   {instances}
  lake at targ1;   {instance name = typename}
  pump1 : pump (rate=30)   **at** sun1;
  reservoir (capacity=2000)   **at** sun 1;
  user1 : user (consumption=4) **at** targ2;

link   {binding of interfaces}
  pump 1.waterin   **to** lake.waterout;
  pump 1.waterout   **to** reservoir.waterin;
  reservoir.pumpon   **to** pump 1.on;
  reservoir.pumpoff   **to** pump 1.off;
  user 1.waterin   **to** reservoir.waterout;
end.

b) Textual Representation

Fig. 4 Configuration Specification for Water Supply System

III.D Hierarchical Composition

Conic supports hierarchical composition in that a group module type may be constructed out of nested instances of task modules or other group modules. Fig. 5 shows that the user module in fact consists of two separate task modules, one which meters water flow and the other consumes water. A group module also has an interface defined in terms of exitports and entryports, and can be parameterised. From the outside it is not possible to distinguish a group module from a task module i.e. the interface to both is described in the same way and they are treated the same for configuratin purposes.

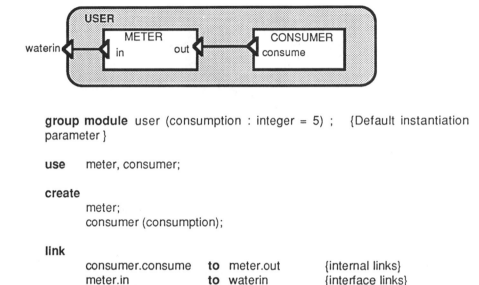

group module user (consumption : integer = 5) ;   {Default instantiation parameter }

use   meter, consumer;

create
        meter;
        consumer (consumption);

link
        consumer.consume   **to** meter.out   {internal links}
        meter.in   **to** waterin   {interface links}
end.

Fig. 5 A Group Module Encapsulates Other Modules

A Group module which contains an executive is called a *logical node*. The task modules within a group module execute concurrently and share an address space. The executive supports multi-tasking, communications and dynamic configuration and allows a node to run as a Unix process or on a bare machine.   The executive is itself a set of Conic modules which can be configured for different hardware using the Conic configuration language [Magee 87].   The logical node is the unit of distribution and dynamic configuration. Fig. 4 is in fact a simplification because the modules lake, pump, reservoir and user would have to be logical nodes containing an executive.

The system is actually constructed by submitting the configuration description to a configuration manager tool which downloads module code into target processors or instantiates processes under Unix as appropriate (see Fig. 6). The configuration management tool and its supporting environment is described in [Magee 87].

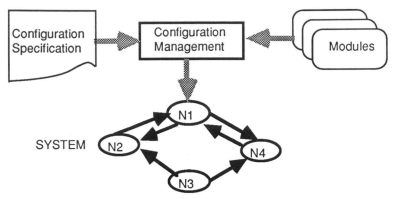

Fig. 6 System Construction

IV DYNAMIC CONFIGURATION

IV.A Requirements

Changes to be applied to a system include modification of a function already provided by the system, or extension by the introduction of new functions. There may be a change in implementation due to new technology, improved implementation techniques or to provide redundancy. In addition to these **evolutionary** changes, the system must cater for **operational** changes such as replacing failed components, moving components to improve performance. It is certainly advantageous if the system is capable of supporting such change dynamically, without interrupting the processing of those parts of the system which are not directly affected.

Change specification is an extension to configuration specification so should also be *declarative* in terms of the *system structure*. System evolution at the level of programming in the small is at too low a level, being both too detailed and unsupportable due to the tight coupling of program elements. Instead, changes should be specified at the configuration level in terms of create/destroy components and link/unlink interfaces. Understanding the effects of change is easier at this level. In addition interfaces are more clearly defined at this level making it easier to determine change dependencies i.e. what other components are affected by a change to one component.

The configuration management support system, and not the programmer, should be responsible for determining the specific ordering of actual change operations applied to the system. This clearly separates the change required (which is application specific) from *how* it is to be executed. Configuration management can exploit parallelism for implementing changes. Declarative specifications leave such decisions to the implementation mechanisms. Another advantage of the declarative approach is that the current configuration can be easily derived from an initial specification plus change specifications (see Fig. 7).

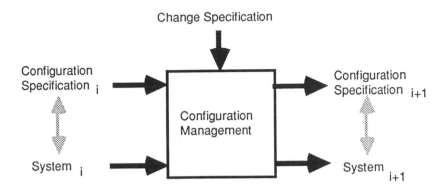

Fig. 7 Dynamic Configuration

IV.B Preserving Consistency

Consistency is usually expressed in terms of some application dependent invariant. Application transactions modify the state of the application, and, while in progress, have transient state distributed in the system. A system is viewed as moving from one consistent state to the next. While transactions are in progress the internal states of nodes may be mutually inconsistent. In order to avoid the loss of application transactions and achieve a consistent state after change, a consistent application state is required before system change. *Maintaining application consistency should be the responsibility of the component programming level.*

Changes should minimise the disruption to the application system. It should not be necessary to stop an entire application system running to modify part of it. The management system should be able to determine a minimal set of nodes which are affected by the change, from the change specification and the initial configuration specification. The rest of the system should be able to continue its execution normally.

When a change specification is applied to a set of components, the configuration manager determines which components are affected. These components are informed that they are about to be changed. It is their responsibility to achieve consistency, which may require some processing and inter-component communication. When the components inform the configuration manager that they have reached a consistent state, the individual operations to change the components can be performed (see Fig. 8).

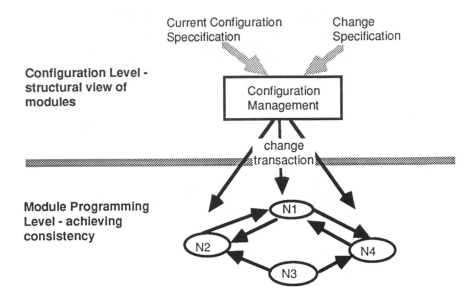

Fig. 8 Change Management Separation of Concerns

The management system does not force changes but waits for the application to reach a consistent state. This consistent state implies that there is no communication in progress between the affected nodes. Each component is said to be in a **quiescent** state. Further, the components must remain quiescent while the change is executed. This gives newly created components the opportunity to be initialised in a state which is consistent with the rest of the system and components which are being destroyed have the opportunity to leave the system in a consistent state.

We have identified a generic interface between applications and configuration management for performing dynamic reconfiguration, as well as a set of configuration states for components. The details of this are beyond the scope of this paper, but more information can be found in [Kramer 90].

IV.C Dynamic Configuration in Conic

We return to the example in section III.A, to illustrate dynamic configuration supported by Conic. Fig. 9 shows an extended water supply system with an additional pump and user.

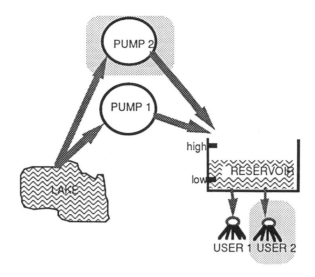

Fig. 9 Extended Water Supply

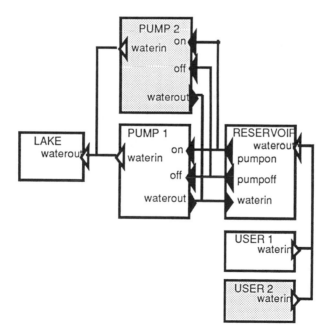

Fig. 10 Graphical Version

Fig. 10 shows the graphical representation of the extended system and the textual change script needed to produce the extended system is shown in Fig. 11.

```
manage  watersystem1

create pump 2 (20) : pump at   sun2
create user 2 (5) : user    at   target3

link    pump 2.waterin      to   lake.waterout
link    pump 2.waterout     to   reservoir.waterin
link    reservoir.pumpon    to   pump 2.on
link    reservoir.pumpoff   to   pump 2.off
link    user 2.waterin      to   reservoir.waterout

start   pump 2
start   user 2
```

Fig. 11 Dynamic Reconfiguration Specification

IV.D Support Facilities

The Conic environment assumes one or more Unix systems are available for software development and initial testing. An application can then be distributed amongst multiple Unix systems and/or bare target computers if real-time response is required. The basic support facilities required at each computer in a distributed system includes:

*   The ability to download code (logical node type) from the development system.
*   Creation of node instances
*   Starting, stopping & destroying node instances
*   Linking and unlinking of node ports.
*   Querying the port types on a node and what they are linked to.
*   Querying what nodes exist in a system and their current status.

Each user communicates with a configuration manager (CM) via a graphical and/or textual user interface. The CM obtains information on names, network addresses, and status of nodes from a registration server. Nodes periodically report status to the server, which can be replicated for reliability. CM links itself to individual nodes to query node ports or up-to-date status, and to perform configuration changes to nodes (see Fig. 12).

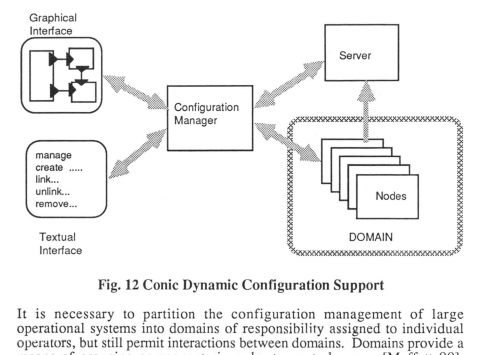

Fig. 12 Conic Dynamic Configuration Support

It is necessary to partition the configuration management of large operational systems into domains of responsibility assigned to individual operators, but still permit interactions between domains. Domains provide a means of grouping components in order to control access [Moffett 90]. Structuring the configuration management of large systems can be achieved by defining subdomains containing subsets of the components, which may correspond to particular subsystems, and assigning these to particular operators. Components can also be members of multiple domains.

V   INTEGRATING HETEROGENEOUS COMPONENTS

In section I, we identified the need to integrate components implemented using various programming languages best suited to the various aspects of an operational system. We now discuss the various approaches to accommodating heterogeneity and explain how this has been achieved using the configuration approach.

V.A Linking Foreign Procedures

It is possible to call procedures written in another language such as C, or Fortran from a Conic component. This only works if the foreign language is a procedural language which is "similar" to Pascal. It requires a standard procedure calling convention in terms of how parameters are passed to and from the procedure, but most compilers in a particular environment (e.g. Sun Unix) already abide by a common calling convention. The foreign procedures must reside in the same address space as the Conic component and so are linked in with the code of the component. It is not possible to call remote procedures on different machines. Another disadvantage is that parameters must have the same representation in both Pascal and the foreign procedure. For example it is not possible to pass a Pascal record to a Fortran procedure.

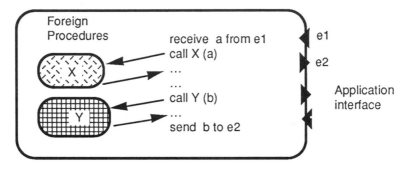

Fig. 13 Foreign Procedures Within a Component

This type of integration of foreign languages is supported by most compilers and linkers. It is effectively transparent to the Conic configuration system as the integration is within a single component (see Fig. 13). The lack of flexibility and the inability to cater for distribution make it inappropriate as the basis of a generalised approach to dealing with heterogeneity.

V.B Shadow Components

Many languages and packages have a built in run time environment which would not be easily modified to incorporate standard communication primitives to enable foreign components implemented with those languages to communicate with the rest of the system. Often the only interaction with the external environment is via standard operating system interfaces such as files or terminals. The solution is to provide a *shadow* process which translates the standard communications interface (i.e. a Conic interface) into local operating system calls e.g. via Unix pipes as shown in Fig. 14.

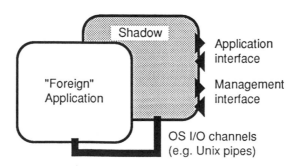

Fig. 14 Shadow Process as an Interface

The shadow process provides both an application interface and an interface for configuration management. The configuration management systems treats the pair of processes as a single indivisible configuration unit.

This approach was used in the Conic environment to integrate Prolog components as we did not have the sources of the Prolog run-time system, so were unable to modify it to directly support Conic communications. The disadvantage of this approach is that file and terminal I/O are not as flexible as the Conic communication primitives and other research groups have modified AI languages to support communication primitives [Jones 85].

V.C Interface Specification Language (ISL)

The ISL provides a common language for specifying the interface to a component in a way that is independent of the programming language used to implement the component [Lamb 87, Hayes 87, 88, Jones 85, Falcone 87].

The minimum information that must be specified is the interaction points. In Conic terms this would be the set of ports on a component. In systems which implement Remote Procedure Calls (RPCs) this would be the set of procedures which form the service offered by a component (exported interface) together with the external procedures called by a component (imported interface). In addition the signature of each interaction point must be specified i.e. the message types or procedure parameters.

In general an ISL includes:

i)   A *data typing language* to define the information transferred across the interface. In Conic, the data typing language was based on Pascal data types with a few extensions. This is used to specify the message types (i.e. records) which can be sent or received by a component.

ii) A means of defining named *Interaction Points* which in Conic terms are the exitports and entryports of an interface.

iii) A specification of the *interaction protocol* to be used.   Conic only supports bidirectional request - reply and unidirectional, asynchronous notify interactions. If there is no reply part to a port then it must be a notify port.

The Conic programming language was defined by extending Pascal to include support for modularity and the ISL. A Pascal compiler was modified to include these extensions. The ISL was incorporated into the configuration language so that an interface to a composite component can also be described using the same ISL.

In addition the Smalltalk run time system was modified to permit a Smalltalk subsystem to present a Conic interface to remote components and to interact with these components using the standard Conic primitives. This was accomplished by defining Smalltalk classes corresponding to Conic ports, messages and tasks. Also Smalltalk task classes to implement the normal Conic run environment had to be provided. A Smalltalk environment running on a workstation corresponds to a Conic logical node but, tasks and ports can be created and deleted dynamically as Smalltalk supports dynamic object creation and deletion.

VI CASE STUDY

This section describes the experience with the Conic toolkit and its application to automation and heterogeneous integration for Process Control at an industrial user's site.

VI.A Background

In the period 1987-91, computer scientists from British Petroleum Research (Sunbury-On-Thames, UK) were investigating application of state-of-the-art information technology (IT) for Process Control. The aim was to exploit the novel features of advanced computer systems, especially in the context of networking environments, and to create a framework for the development of multi-disciplinary techniques. The planned outcome was to form a test bed for research in chemistry, thermodynamics, process modelling, advanced control algorithms and novel IT techniques.

The Information Science and Engineering team selected the Conic toolkit to develop three distributed real-time applications on two different pilot plant installations in their laboratories at the Sunbury Research Centre. One application was for a Batch Distillation Column and two different applications used a Chemical Reactor plant.

The main rationale behind the choice of Conic was the apparent match between the distributed paradigm and the application domain, lack of any sophisticated commercial distributed programming environment, perceived benefits of dynamic reconfiguration and potential for heterogeneous language integration. In addition the Conic development system allowed cross-compilation for VME Bus target platforms, which performed efficient and cost-effective handling of special I/O functions in a Process Control system.

VI.B Building Up Conic Expertise

The first two systems deployed at Sunbury Research Centre were intended to gain expertise in the use of the Conic environment. The team of users of the Conic toolkit found it easy to learn and intuitive to apply. They appreciated its features that made it useful for building real-time, distributed programs, as well as for the operational management of the whole software installation.

The first Conic installation was the Batch Distillation Column computer system. It was developed as a prototype environment, where components written in different languages handled specific control, monitoring and simulation tasks. The three approaches to the integration of heterogeneous components were applied to achieve the harmonious cooperation of multiple run-time environments:

* use of the ISL method to develop a Smalltalk user interface;
* the Shadow Component method for the integration of a rule-based decision making component;
* use of foreign procedures, to Fortran and C routines for the on-line simulation and modelling.

The system was run alongside a standard commercial control system. The successful handling of the plant data and its economics have been illustrated in operation.

After the Batch Distillation Column work was completed, the work switched to the Chemical Reactor installation. The first Conic system enabled the integration of an advanced control algorithm for temperature control of exothermic reactions. The Conic software provided the implementation of the basic functions in the system: distributed data acquisition, monitoring, logging and performed the advanced control algorithm. An existing HP300 computer was included as a subsystem of the overall configuration.

VI.C   Chemical Reactor Pilot Plant

In 1990 the reactor was redesigned and the Conic system was to take over the full control of the plant. The purpose of this new application was to experiment with fundamentals of a specific chemical reaction and the computer system was to automate certain tasks. The chemists supplied a functional specification which the new system architecture had to meet.

The reaction, details of which are confidential, was between 3 gas feeds and a liquid, in the presence of a catalyst. Each feed had a flow regulator which controlled the flow of the reagents. The walls of the reactor were heated by 4 heaters and the core temperatures were read from 6 thermocouples. The pressure in the reactor was maintained with a pressure regulator.

The basic instrumentation was done with standard flow, temperature and pressure controllers (Eurotherm, Honeywell, etc.), wired up to a PPM device, which in turn communicated with an IEEE488 gateway computer (HP300). The output of the reactor was analysed with a mass spectrometer, run by a stand-alone PC. In addition a computer controlled HPLC pump was used.

The Conic system runs on one Sun3 workstation and three 68020/VMEbus target computers, linked by Ethernet. Fig. 15 shows the basic layout of the reactor.

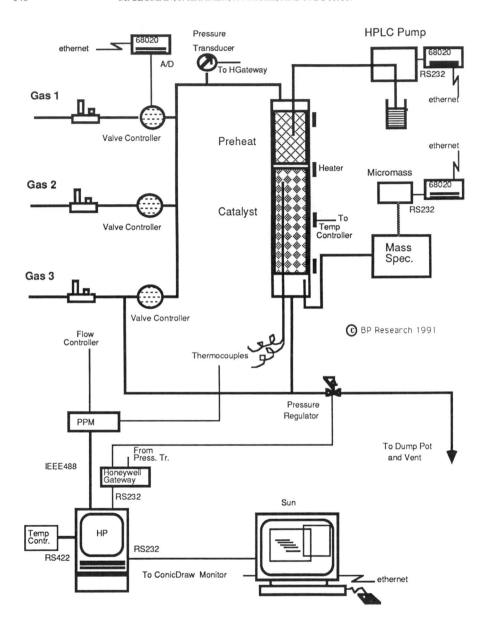

Fig. 15 Pilot Plant Reactor

The main requirements that the designers of the system had to meet could be summarised as follows:

1. Provide a control system with standard data acquisition, monitoring and logging functions, which is robust and usable enough for a Process Control environment

2. Automate scheduling algorithm as required by the reaction definition

3. Allow real-time response for certain time critical operations (eg. synchronisation of events, handling of complex data, such as the mass spectrometer interface)

4. Enable application of multi-disciplinary techniques, by the use of highly specialised functional components, possibly having their own run-time system

5. Integrate existing IEEE488 subsystem, providing interface to flow controllers, wall heater controllers, core temperature sensors and the pressure regulator via an HP300 computer. Reuse previously developed modules where applicable

6. Allow for future extensions to accommodate further novel components

7. Supply easy-to-use operator interface, based on a standard look-and-feel graphical user interface

8. Make provision for operational failures and automatic failure recovery

VI.D   The Conic Solution

It was apparent that only a distributed system was capable of satisfying all these requirements. The Conic distributed paradigm provided the necessary model on which the construction of real-time system could be based. Conic message passing primitives allowed the synchronisation and communication of concurrent components and dynamic reconfiguration made it suitable for fault tolerance and incremental development.

The system was functionally decomposed into a set of concurrent components and their communication arrangements were designed using the graphical tool, ConicDraw [Kramer 89]. The result was the fully distributed application illustrated in Fig. 6.2 (ConicDraw is normally used to monitor operational systems). It is a client/server architecture, shown as a network of nodes with their ports and run-time links.

The following Conic components provided interfaces to the hardware resources i.e. sensors and actuators and performed the role of server components:

ppm - provides access to instruments and sensors via the HP gateway (PPM);

micromass - handling the serial output from the Mass Spectrometer PC (Micromass System);

pump - serial line interface to the HPLC pump;

valves - managing the flows of gas reagents.

The following components performed application specific functions and took the role of clients.

logger - logging in asynchronous mode (pump change, flow change) and synchronous mode (sampling of temperatures and pressure);

pacemaker - scheduler, coordinating the logic of the experiment;

operator interface - interactive control, monitoring, experiment definition and procedural activation (eg. logging, start experiment, stop experiment). It was based on an shadow component: an X- protocol client

valvesun - interactive control over Valve hardware from the Sun. It was based on a shadow component: an X- protocol client

The initial system consisted of a Sun based graphical interface to control valves on an experimental rig (i.e. components *valvesun* and *valves*). The other components were gradually implemented and added to the system to reach the current configuration shown in Fig. 16.

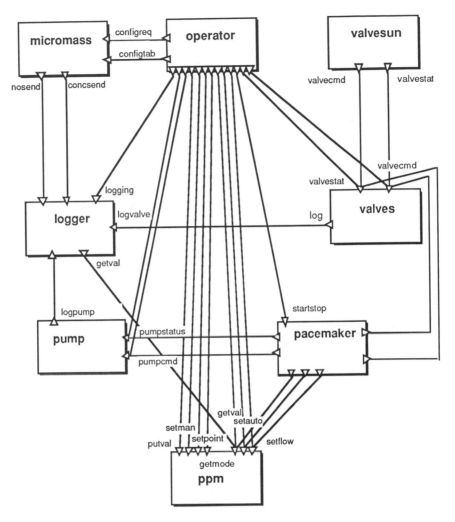

Fig. 16 Conic Configuration for the Reactor

VI.E Evaluation

The advantages of using the Conic toolkit to develop the system can be explained by analysing how the requirements of the application were satisfied.

1) Standard Control System

The system was implemented with Conic nodes encapsulating the hardware resources of the system and performing the role of servers for real-time data.. This allowed the client nodes to read and manipulate the process variables and perform the control, monitoring and logging functions.

2) Automation

Automated tasks were specified using an Open Look (Xview) graphical user interface. The details of the experiment were generated to a set of files and a separate Conic node used those files as the basis for a control sequence, executed when required.

3) Real Time Response

Real-time response was accomplished by:

* Running concurrent components on different machines, with synchronisation and communication using the Conic message passing primitives.
* Use of the internal concurrency for multiple tasks in a Conic node to delegate the slow I/O functions to tasks which waited for their completion - eg. continuation of next task while previous task waits for the confirmation of a remote request, when 3 feeds were manipulated simultaneously.
* Withdrawing from failed transactions (no hold-ups).
* Use of asynchronous sends and selective receives for requests from different sources, as the basis for efficient communication.

4) Heterogeneous Integration

Both calls to foreign language procedures (C) and shadow components were used in this application. There were two different modes in which Shadow Components were used:

i) The master mode: where the Conic Shadow process is acting as the parent of the foreign process (usual for a functional node - here intended for some additional components).

ii) the slave mode: where the foreign process forks the Conic shadow process, thus gaining access to the message-passing capability, but essentially being driven by its own events (eg. the Operator's X-Window interface and its Shadow Conic node).

5) Integration of an Existing Installation

It was due to the inflexibility and the limited power of the HP subsystem that the need for a more powerful, flexible and scalable computer system had created an opportunity to develop a new system for this particular rig.

The integration was achieved by providing a front-end node that handled all communications with the HP IEEE488 gateway via requests over a serial link and communicated with the rest of the Conic system via messages. This was reused, with some modification, from the previous experiment.

6) Expandability

The incremental system development meant that a small subsystem was constructed first as a proof of concept, for the designers, and to illustrate an important function, for the benefit of the user. The application chosen was the use of graphical interfaces for the remote operation of valve hardware. This approach resulted in gain in confidence by all involved and satisfied the fitness-for-purpose requirement.

Both the control system and the scope of the experimental work are being extended by applying advanced control algorithms, developed previously. Additional development is taking place to include novel components within the live system (eg. a knowledge-based real-time component, a neural network controller, an on-line parallel model). Again, it will be the heterogeneous integration feature of Conic that will facilitate these extensions.

7) Standard Graphical User Interface

This was possible due to the combination of standard X-Window application with a Conic Shadow component. It resulted in the interface being built as a graphical desktop, with separate windows and icons for various Sun-based nodes. Users of the system easily adapted to using this interface e.g. the mouse driven menus for start-up and recovery, sliders for interactive manual control, etc. The whole experiment could be started up and managed within this environment.

8) Failure Recovery

At the component programming level, the Conic programming language provided the capability to detect failed remote operations and take recovery action. At the configuration level, the Conic Configuration Manager permitted the recovery of failed nodes when the component level recovery was not implemented or insufficient. Automatic routines have been designed that helped to recover from node failures and restore communication channels (links).

VII CURRENT WORK

The Conic environment has been supplied to about 25 institutions around the world and it has been used for both research and teaching. Example applications include reconfigurable control, object oriented run time support system, distributed simulation, object oriented database support, robotics control, telecommunications control and simulation etc. Students can learn the language very easily if they have a knowledge of Pascal or

Modula 2. It provides a powerful tool for programming concurrent systems whether they are distributed or within a single machine.

Our experience with Conic led us to develop a new ISL and Configuration Language. This work was started within an SERC ACME project in collaboration with BP and continued as part of an Esprit funded Collaborative project called REX - Reconfigurable and Extensible Parallel and Distributed Systems [Magee 90, 91]. Rex overcomes some of the problems identified by users of Conic. For example Conic configuration specifications can be rather verbose, as a link statement is needed for each port in the interface. A REX interface is defined as a set of ports and then a single link statement is used to link all ports in the set to the corresponding port set on another component. The interface is specified independently from the component template which implements that interface, thus allowing multiple different implementations for an interface. Also the same interface definition can be used to generate exitports for client interfaces and entryports for server interfaces.

The Conic group module could only reside in a single node whereas the REX composite component can be distributed across multiple phsyical nodes. REX also provides a means for specifying preplanned programmed changes at the configuration level and invoking them from within a running program component.

The REX ISL and Configuration language are programming language indpendent, whereas Conic was rather Pascal oriented. Preprocessors are being developed for C++, Modula 2 and Prolog. In addition tools are being devleoped to support the design and analysis of distributed systems.

There is a need for more flexible communication primitives. For example scheduling servers may not respond to requests in the order in which they were received and there should be more direct support for remote procedure call type of interaction as these are becoming very common in many systems.

VIII CONCLUSIONS

The system designers/developers at BP and their customers identified the following benefits in their use of Conic:

-   Conic allowed the demonstration of how distributed applications can be built and showed that they meet design and operational requirements that conventional systems could not meet.

-   The concepts supported by Conic provided a good paradigm for the modelling of a real-time distributed application.

-   Dynamic Reconfiguration was used to incrementally develop the initial system, and later for managing the distributed computer installation. This increased the availability and allowed further expansion, as system structure evolved to meet new requirements;

- The Conic system served as a cohesive framework for integration, multidisciplinary technology and groupwork in a heterogeneous environment.

- The benefit to the customers of the system were: the automation of plant procedures, intelligent scheduling and control, and facilities for multi-disciplinary team research, which stimulated work and enhanced the scope of the application.

The paper has outlined the principles relating to configuration of software for operational systems. A configuration is specified in terms of the structural relationship of component instances.

1. *The configuration language used for structural description should be separate from the programming language used for basic component programming.*

   This was the key feature of both Conic the current work in REX and permits the construction of systems which consist of heterogeneously programmed components.

2. *Components should be defined as context independent types with well defined interfaces.*

   The Conic ISL specifies the ports through which external services are accessed as well as the ports which represent the service that a component provides. Both are needed to define behavior of reuseable components. A component thus makes no direct references to non-local entities. It can thus be used in many different contexts by binding its ports.

3. *Using the configuration language, complex components should be definable as a composition of instances of component types.*

   This is one of the most powerful feature of the Conic configuration languages as it permits a complex subsystem to be treated as a single component for configuration purposes. Both primitive and composite components are treated identically for specifying interfaces and configuration.

4. *Change should be expressed at the configuration level, as changes of the component instances and/or their connection.*

   The fact that both initial configuration and subsequent changes are specified at the configuration level makes it much easier to determine the current configuration of a system. This is essential for management purposes, as it forms a specification of what should be in the system. It also permits dependencies to be determined.

ACKNOWLEDGEMENTS

Acknowledgement is made to our colleagues Essie Cheung, Steve Crane, Naranker Dulay, Anthony Finkelstein, Keng Ng, Kevin Twidle at Imperial College, John Haberfield, Trevor Lilley and John Sharp at BP. We gratefully acknowledge the SERC ACME Directorate under grant GE/E/62394, the SERC under grant GE/F/04605 and the CEC in the REX Project (2080) for their financial support.

REFERENCES

ANSA 89   Ansa Reference Manual, Available from Ansa, 24 Hills Road, Cambridge, CB2 1JP (Email: apm@ansa.co.uk).

Barbacci 88 M.R.Barbacci, C.B.Weinstock, J.M.Wing, "Programming at the Processor-Memory-Switch Level", *Proc. IEEE 10th Int. Conf. on Software Eng.*, Singapore, April 1988.

Birrell 84   A.D. Birrell, B.J. Nelson, Implementing Remote Procedure Calls, *ACM TOCS*, Vol. 2(1), Feb. 84, pp. 39-59

Bloom 83   T.Bloom, "Dynamic Module Replacement in a Distributed System", Technical Report MIT/LCS/TR-303, MIT Laboratory for Computer Science, March 1983.

deRemer 75 F.deRemer, H.Kron, "Programming-in-the-large versus Programming-in-the-small", *Proc. Conf. Reliable Software*, 1975, pp. 114-121.

Coulas 87   M. Coulas, G. Macewen, G. Marquis, "Rnet: A Hard Real-time Distributed Programming System", *IEEE Trans. on Computers*, C13(8), Aug. 1987, pp. 917-932.

Friedberg 87 S.A.Friedberg, "Transparent Reconfiguration requires a Third-Party Connect", TR220, Computer Science Department, University of Rochester, New York, Nov. 1987.

Goguen 86 J. Goguen, "Reusing and Interconnecting Software Components" *IEEE Computer,* Vol 19(2), Feb. 1986.

Falcone 87 J. Falcone *A Programmable Interface Language for Heterogeneous Distributed Systems*, ACM TOCS 5(4) Nov. 1987 pp. 330-351

Hayes 87   R. Hayes, R. Schlichting, *Facilitating Mixed Language Programming in Distributed Systems*, IEEE Trans. on Software Eng., SE 13:12, Aug, 1987.

Hayes 88   R. Hayes, S. Manweiler, R. Schlichting, *A Simple Systems for Constructing Distributed Mixed Language Programs,* Software Practice and Experience, 18(7), July 1988, pp.641-660.

Jones 85   M.B. Jones, R. Rashid, M. Thompson, "Matchmaker: An Interface specification Language for Distributed Processing", *Proc. 12th. Annual Symp. on Principles of Programming Languages,* Jan. 1985, pp. 225-235

Kramer 85   J.Kramer, J.Magee, "Dynamic Configuration for Distributed Systems", *IEEE Transactions on Software Engineering*, SE-11 (4), April 1985, pp. 424-436.

Kramer 87   J. Kramer, J. Magee, M. Sloman, "The CONIC Toolkit for Building Distributed Systems", *IEE Proceedings*, Vol. 134, Pt. D(.2), March 1987.

Kramer 89   J. Kramer, J. Magee, K. Ng, "Graphical Configuration Programming", *IEEE Computer,* 22(10), October 1989, 53-65.

Kramer 90a   J. Kramer, J. Magee, A. Finkelstein, "A Constructive Approach to the Design of Distributed Systems", *10th Int. Conf. on Distributed Computing Systems,* Paris, May 1990.pp. 580-587.

Kramer 90b   J. Kramer, J. Magee, "The Evolving Philosophers Problem: Dynamic Change Management", *IEEE Transactions on Software Engineering,* SE-16 (11), Nov. 1990, pp. 1293-1306.

Kramer 91   J. Kramer, J. Magee, M. Sloman, N. Dulay, " Configuring Object-based Distributed Programs", May 1991.

LeBlanc 85   T.J.LeBlanc and S.A.Friedberg, "HPC: A Model of Structure and Change in Distributed Systems", *IEEE Trans. on Computers*, Vol. C-34 (12), December 1985.

Magee 89   J.Magee, J.Kramer, M.Sloman, "Constructing Distributed Systems in Conic", *IEEE Transactions. on Software Engineering*.SE-15 (6), June 1989, pp. 663-675.

Magee 90   J.Magee, J.Kramer, M.Sloman, N. Dulay, "An Overview of the REX Sofware Architecture", *IEEE Computer Society Workshop on Future Trends of Distributed Computing Systems,* Cairo Oct. 1990, pp. 396-402

Moffett 90   Moffett J.D. Sloman M.S. & Twidle K.P., Specifying Discretionary Access Control Policy for Distributed Systems, *Computer Communications*, vol 13 (9), pp 571-580 (November 1990).

Sloman 89   Sloman M.S. & Moffett J.D., Domain Management for Distributed Systems, in Meandzija & Westcott (eds), *Proc. of the IFIP Symposium on Integrated Network Management,* Boston, USA, May 1989, North Holland, pp 505-516.

Snyder 86   A. Snyder. "Encapsulation and Inheritance in Object-Oriented Programming Languages", *OOPSLA '86 Proceedings*, pp.38-45.

Schwan 86   K. Schwan, A. Jones, "Flexible Software Development for Multiple Computer Systems", *IEEE Transactions. on Software Engineering,* SE12(3), March 1986, pp. 385-401.

Wegner 90   P.Wegner, "Concepts and Paradigms of Object-Oriented Programming", *OOPS Messenger (ACM SIGPLAN)*, Vol. 1 (1), August 1990, 7-87.

SOFTWARE FOR DYNAMIC TOOL MODELING IN MANUFACTURING AND AUTOMATION

LUIS C. CATTANI
PAUL J. EAGLE

Department of Mechanical Engineering
University of Detroit Mercy
Detroit, Michigan 48221

I.   INTRODUCTION

The number of robot applications has increased dramatically since 1980 and the predominant application has been spot welding in the automotive industry. Unfortunately, spot welding tools or guns typically impart large loading on robots and have made robot reliability an important issue. Mechanical failures have been identified with the symptoms of excessive backlash and harmonic drive failures. Excessive backlash can cause poor weld quality while the harmonic drive failures can cause downtime [1].

In spot welding applications, robot overloading is caused by the weld guns producing large reaction torques on the robot, exceeding the robot design capacity. An often overlooked source of robot loading occurs due to dynamic effects. This is due to the difficulty of calculating the moments of inertia of the welding guns. This complex geometry is a result of the sizes and shapes of the welding tools needed to operate on car bodies.

To address this problem, a simple solid modeling software called **TANGO-UD** has been developed to calculate and analyze solid properties (mass, volume, moments of inertia, products of inertia and radius of gyration) of welding guns and other mechanical components. The software has a geometric and material definition language, a parser and compiler, a mass properties calculator, a mass properties analyzer and the ability to display the wire frame model of the object being analyzed using the **AutoCAD** graphics software.

The software is based on a **primitive-volumes** method for modeling the welding gun shapes. A designer can quickly build up complicated shapes by combining volume elements such as a right parallelepiped (**BOX**), a wedge (**WED**), a cylinder (**CYL**) and a cylinder sector (**SEC**). Any shape can be rotated and translated with respect to a specified coordinate frame. Negative volumes can be defined to denote an absence of material, such as a conduits for air or water passages. A reference system can be positioned and oriented with respect to the global frame for referring to the coordinate system of the robot drive that is being loaded by the welding gun. This approach is particularly useful because the inertia tensor of the gun is later calculated with respect to this system that is often remote from both the global frame and the centroids of the gun. Ultimately, the software generates a mass properties report that is useful for design engineers in the analysis of welding gun dynamics.

Because the method uses only four primitives, the software requires surfaces to be modelled as planes and parts of cylinders. Furthermore, the program does not have high-quality visual effects as many of the academic and industrial solid modeling packages. However, the efficiency of the program and the simplicity of the description language are the relevant issues of TANGO-UD when a complex engineering problem must be solved.

II. SOLID REPRESENTATION METHODS

Computer-based systems for modeling the geometry of rigid solid objects do not manipulate physical solids, instead they manipulate symbol structures which represent solids. It is the existence of these abstract models that allows us to study mathematically, without resorting to physical experiments, the properties of bodies such as volume and mass.

Some of the methods for representing solids with the essential mathematical properties such as closure under boolean operations, finite describability and boundary determinism [2] are:

> **Primitive Instancing**
> **Quasidisjoint Decomposition**
> **Simple Sweeping**
> **Boundary Representation**
> **Constructive Solid Geometry**

These methods have many inherent advantages and disadvantages related to their ease of implementation, computational overhead, accuracy of modeling and ease of use for a designer. Many solid modeling packages described in the literature utilize these techniques [3 - 12].

Many common solid objects have shapes characterized by surfaces which are composed of large numbers of simple faces. Most mechanical assemblies and machined components are examples of this class. They may have hundreds of faces, but most faces will be planar, cylindrical, or some other elementary surface form. Robot welding tools or weld guns have very complex three-dimensional geometry, however, their shape could be approximated by combining primitive volumes. TANGO-UD uses four primitives [13-15]: **BOX** (right parallelpiped), **WED** (wedge), **CYL** (cylinder) and **SEC** (sector of a cylinder) to describe the geometry of the welding tool by adding together or subtracting these simple primitives (Figure 1).

This scheme has three main advantages:

a.   The quantity of input is less than if faces, edges and their interconnections were input separately.

b.   Practicing designers prefer to think of objects as volumes of material enclosing a given void rather than faces and edges.

c.   Solid properties for each primitive component of the object can be combined (added and subtracted) to create complex composite objects with a low computational cost.

The disadvantage of this approach is that the efficiency of the inertial properties calculations will depend on the approximation effects introduced by modeling a complex geometry by only four primitives. It is the responsibility of the user to choose the appropriate primitives for building a model. Even with this restriction of four primitives and surfaces limited to planes and parts of cylinders, the variety of objects which can be approximated and dynamically analyzed by TANGO-UD covers a large percentage of the components found in a typical weld gun. On the other hand, the simple geometric definition language used to drive the program seems appropriate for mechanical designers. TANGO-UD is at a level of complexity comparable to the APT Numerical Control language that has been widely used by machine part programmers.

III.   GEOMETRIC FORMULATION OF THE TANGO-UD SYSTEM
A primitive instance has been chosen as a representation scheme for the geometric modeller. The mathematical forms and data structures suitable for this approach must be implemented such that they provide necessary

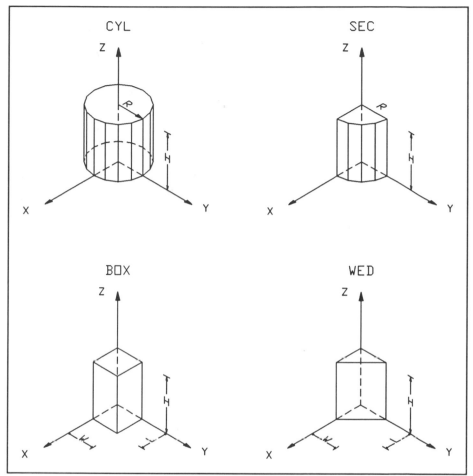

Figure 1 TANGO-UD Primitives

functions in a solid modeling system:

a.   A useful set of primitives used to build complex objects.

b.   A complete set of geometric tools for building consistent objects and space relationships (translation and rotation).

c.   Display of a wire-frame or other format picture to visualize the modelled object.

d.   Computation of mass properties.

A definition language is the tool used by a designer to input primitives, dimensions and spatial relationships among primitives. TANGO-UD provides for spatial relationships among primitives using symbolic operators. The main idea is that each primitive component of an object has an embedded coordinate frame which is defined relative to the world frame. The world frame is a coordinate frame that is used as the global system of the object, translations and rotations of primitives are with respect to this system. The location of a primitive can be transformed to anywhere in the world system, but the locations of the embedded axes remain the same within the primitive. A third frame, called the reference frame, is defined relative to the world system and is used as the base system for the calculation of the inertia tensor of the complex object and its primitive components.

This representation is attractive since after the geometry of the weld gun or other object has been modelled, the inertia tensor (proportional to driving torques of the gun) can be calculated for different positions and orientations of the reference system, representing the probable positions and orientations of the robot drive motors.

Frames can be expressed by the homogeneous transform [16], a 4 x 4 matrix containing orientation and position information. The matrix contains an orientation submatrix, ${}^{A}_{B}R$ and a position submatrix, ${}^{A}P_{B_{ORIGIN}}$. The notation is interpreted as the matrix or transformation ${}^{A}_{B}T$ describing the frame {B} relative to the frame {A}. The columns of ${}^{A}_{B}R$ are unit vectors defining the directions of the principal axes of {B} respect to {A}, and ${}^{A}P_{B_{ORIGIN}}$ locates the position of the origin of {B} relative to {A}:

$$\begin{array}{c} {}^{A}_{B}T = \left[\begin{array}{ccc|c} & {}^{A}_{B}R_{3\times3} & & {}^{A}P_{B_{ORIGIN}3\times1} \\ - & - & - & - & - & - \\ 0 & 0 & 0 & 1 \end{array} \right] \end{array} \qquad (1)$$

Transformations can be expressed using the same homogeneous transformation matrices. A pure translational matrix is represented by $Trans(x,y,z)$, and pure rotations about x-axis, y-axis, z-axis can be denoted by the orthonormal matrices $Rot(x,\theta)$, $Rot(y,\theta)$ and $Rot(z,\theta)$, respectively. Following are their mathematical expressions:

$$Trans(x,y,z) = \left[\begin{array}{cccc} 1 & 0 & 0 & x \\ 0 & 1 & 0 & y \\ 0 & 0 & 1 & z \\ 0 & 0 & 0 & 1 \end{array} \right] \qquad (2)$$

$$Rot(x,\theta) = \left[\begin{array}{cccc} 1 & 0 & 0 & 0 \\ 0 & \cos\theta & -\sin\theta & 0 \\ 0 & \sin\theta & \cos\theta & 0 \\ 0 & 0 & 0 & 1 \end{array} \right] \qquad (3)$$

$$Rot(y,\theta) = \left[\begin{array}{cccc} \cos\theta & 0 & \sin\theta & 0 \\ 0 & 1 & 0 & 0 \\ -\sin\theta & 0 & \cos\theta & 0 \\ 0 & 0 & 0 & 1 \end{array} \right] \qquad (4)$$

$$Rot(z,\theta) = \begin{bmatrix} \cos\theta & -\sin\theta & 0 & 0 \\ \sin\theta & \cos\theta & 0 & 0 \\ 0 & 0 & 1 & 0 \\ 0 & 0 & 0 & 1 \end{bmatrix} \tag{5}$$

By multiplying a frame by a rotation and/or translation matrix, we can transform the old frame into a new frame.

Mathematical expressions are used to describe the abstract concept of spatial relationships. TANGO-UD describes the structure of the space by a transform equation (Figure 2):

$$\prescript{W}{R}{T}\,\prescript{R}{C_i}{T} = \prescript{W}{P_i}{T}\,\prescript{P_i}{C_i}{T} \tag{6}$$

where:

$\prescript{W}{R}{T}$ describes the reference coordinate system relative to the world frame (this is input by the user),

$\prescript{R}{C_i}{T}$ describes the location of the centroid of the primitive relative to the reference system (this transformation is calculated by TANGO-UD for each primitive),

$\prescript{W}{P_i}{T}$ describes the location of the primitive i relative to the world coordinate system (this is also defined by the user) and

$\prescript{P_i}{C_i}{T}$ describes the centroid of the primitive with respect to its own coordinate system (this matrix is not defined by the user, but is known by TANGO-UD).

Equation (6) shows that it is possible to define the centroid of the primitive relative to the world system in two different ways. This property is used to solve for the unknown transformation that describes the transformation from the centroid of a primitive to the reference system:

$$\begin{matrix} R \\ {}_{C_i}T \end{matrix} = \begin{matrix} W \\ {}_{R}T^{-1} \end{matrix} \begin{matrix} W \\ {}_{P_i}T \end{matrix} \begin{matrix} P_i \\ {}_{C_i}T \end{matrix} \tag{7}$$

This matrix will be used to transform the inertia tensor of a primitive, from its centroid to a reference system.

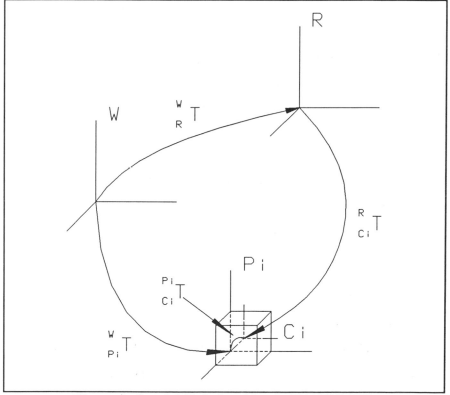

Figure 2 Spatial Relationships in TANGO-UD

The frame information is codified by the TANGO-UD definition language. The statements in the definition language are expanded by the software into an internal form that contains more explicit shape information needed for applications. The information associated with the components of a surface (face, edge and vertex) can be considered in two parts. One is the geometry, including the physical dimensions and location in space of each component. The other is the topology, describing the connections between the components. It should be noted that this use of topology is somewhat different from its conventional use in mathematics. In this case, topology regards a point as vertex that bounds a line to define an edge. Similarly, a ring of edges provides bounds on a surface to define a face. Both geometry and topology are necessary for a complete shape description.

The designer is provided with a set of four shapes or building blocks. These are the basic units of shape and are termed primitive volumes or primitives. They are **BOX** (right-parallelepiped), **WED** (wedge), **CYL** (cylinder) and **SEC** (sector of a cylinder). The shapes are parameterized in such a way that a particular primitive is specified by a few parameters from which TANGO-UD computes a larger, sufficient set of shape information. BOX and WED are parameterized by width, height, length and density. CYL and SEC are parameterized by radius, height and density (Figure 3).

These four primitives are represented (using AutoCAD as a visual communicator) by convex polyhedrons of six faces, five faces, eighteen faces and eight faces for BOX, WED, CYL and SEC, respectively. **Faces** are represented by four points (vertices) that has been previously transformed from the embedded coordinate system associated with the primitive to the world coordinate

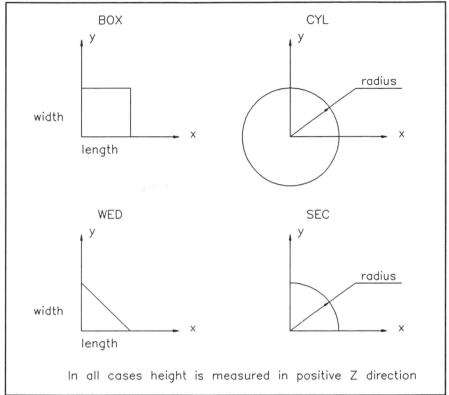

Figure 3 Primitive Parameters in TANGO-UD

system by the transformation $^W_{P_i}T$:

$$^W\bar{p}_{ijk} = {}^W_{P_i}T \; {}^{P_i}\bar{p}_{ijk} \tag{8}$$

where:
$^{P_i}_W\bar{p}_{ijk}$ is the location of vertex *ijk* relative to the primitive frame and $\bar{p}_{ijk}$ is the location of vertex *ijk* relative to the world frame.

The indices can be interpreted as: *i* representing the primitive,*j* representing the vertex number of the surface and *k* representing the surface number (e.g. *k* = 1,...,18 for CYL, where there is one top face, one bottom face and four faces for each quadrant).

Negative shapes, which denote an absence of material, allow hollow regions to be defined. The addition of a positive and a negative volume has the effect of subtracting material from the positive object. Negative shapes are indicated by a negative sign in the density parameter of the primitive.

Representing geometric information is only part of the modeling 3D shapes. As with geometric data, storing all of the relationships between faces, edges and vertices is highly redundant. In fact, one type of relationship is sufficient and all others can be derived from it. TANGO-UD uses the relationship between faces and vertices to define how vertices are joined. This relationship that defines the connectivity between vertices and faces is denoted f:{v}, where f is a set representing the face of a primitive and v represents the vertices of the elements in the set f [12]. Each face is defined by counterclockwise sequences of cartesian points (vertices). The faces are connected in counterclockwise sequences to build the primitive. This topology allows unit normals to each face to point to the exterior of the primitive. Negative primitives could be built with their unit normals to each face reversed, so that they face towards the center of the primitive, permitting a practical method to distinguish between positive and negative volumes in the drawing database.

The data structure (geometry and topology) of the BOX is described below. Analogous structures are used by TANGO-UD for WED, CYL and SEC.

$$BOX_i = (FACE_1, FACE_2, \cdots, FACE_k, \cdots, FACE_6) \tag{9}$$

with

$$FACE_k = (\; {}^W\bar{p}_{i1k}, \; {}^W\bar{p}_{i2k}, \; {}^W\bar{p}_{i3k}, \; {}^W\bar{p}_{i4k})$$ (10)

$$ {}^W\bar{p}_{ijk} = \; {}^W_{P_i}T \; {}^{P_i}\bar{p}_{ijk} $$ (11)

and

$$ {}^{P_i}\bar{p}_{ijk} = \begin{bmatrix} {}^{P_i}p_{ijk_x} \\ {}^{P_i}p_{ijk_y} \\ {}^{P_i}p_{ijk_z} \end{bmatrix} $$ (12)

IV.   MASS PROPERTIES FORMULATION OF THE TANGO-UD SYSTEM

The mass distribution of a 3D body can be characterized by the inertia tensor, which can be considered as a generalization of the scalar moment of inertia of an object. Inertia tensors may be defined relative to any frame in a matrix form. The inertia tensor of a primitive relative to its centroidal frame {C} is expressed as [16, 17, 18]:

$$ {}^C I = \begin{bmatrix} I_{xx} & -I_{xy} & -I_{xz} \\ -I_{xy} & I_{yy} & -I_{yz} \\ -I_{xz} & -I_{yz} & I_{zz} \end{bmatrix}_{3 \times 3} $$ (13)

where the six scalar elements are given by their familiar volumetric integral expressions. The diagonal elements are typically referred to as the mass moments of inertia. The off-diagonal terms are commonly known as the mass products of inertia.

Mass moments and products of inertia for the centroidal inertia tensor, can be calculated for the four primitives (BOX, WED, CYL, and SEC) by standard closed-form expressions. However, the interest is centered on the inertia tensor for the reference system. Thus, transformational properties of the inertia tensor must be considered. Assume that the location of the reference coordinate system $\{R\}$ and centroidal frame $\{C_i\}$ of a primitive are coincident, but have different orientations. If the six inertia terms relative to frame $\{C_i\}$ are known, then the issue is the calculation of the inertia tensor of the primitive relative to the reference frame $\{R\}$:

$$
{}^{R}I = \begin{bmatrix} I_{x'x'} & -I_{x'y'} & -I_{x'z'} \\ -I_{x'y'} & I_{y'y'} & -I_{y'z'} \\ -I_{x'z'} & -I_{y'z'} & I_{z'z'} \end{bmatrix} \tag{14}
$$

The primed coordinates are used to represent a reference coordinate system that is different than the centroidal coordinate system. The six terms of inertia matrix in the above equation can be computed by transformating ${}^{C}I$ into the new frame $\{R\}$ by using a matrix expression in terms of ${}^{R}_{C_i}R$ [17, 18].

The inertia quantities can also be computed relative to a reference system $\{R\}$ that is displaced by a pure translation from the frame at the center of mass $\{C_i\}$ of the primitive. The parallel axis theorem is employed to alter the inertia tensor under translations of the reference coordinate system. The parallel axis theorem relates the inertia tensor in a frame with origin at the center of mass $\{C_i\}$ to the inertia tensor defined with respect to the reference frame $\{R\}$.

The ultimate result of these transformations is the opportunity to compute the inertia tensor of a primitive, with respect to a reference frame $\{R\}$ displaced from the center of mass of the primitive, under translation and rotation. Equation (14) can be used in conjunction with the parallel axis theorem to obtain:

$$^{R}I_i = \begin{bmatrix} I_{x'x'}+m(Y_c^2+Z_c^2) & -(I_{x'y'}+m(X_c Y_c)) & -(I_{x'z'}+m(X_c Z_c)) \\ -(I_{x'y'}+m(X_c Y_c)) & I_{y'y'}+m(X_c^2+Z_c^2) & -(I_{y'z'}+m(Y_c Z_c)) \\ -(I_{x'z'}+m(X_c Z_c)) & -(I_{y'z'}+m(Y_c Z_c)) & I_{z'z'}+m(X_c^2+Y_c^2) \end{bmatrix} \quad (15)$$

The inertia tensor $^{R}I_i$ is calculated by TANGO-UD for each primitive, and the inertia tensor of the aggregate body is obtained by simple matrix additions:

$$^{R}I = \sum_{i=1}^{N} {}^{R}I_i \quad (16)$$

where ^{R}I is the inertia tensor of the complex 3D body relative to the frame $\{R\}$ and N is the number of primitives necessary to construct the body.

V.   TANGO-UD ENVIRONMENT

A typical application of TANGO-UD will be presented as an imaginary design sequence that might be performed. The user knows that TANGO-UD has four alternatives in its main menu: **EDITOR, COMPILER,**

PROMASA, ANALYZER and the program itself has a command language to design shapes by adding or subtracting primitives.

The designers will write the geometric and mass property definition of the object in the TANGO-UD LANGUAGE using an ASCII editor. This definition code will be stored in an input file. The second step is to check the syntax of the input file. This task is accomplished by TANGO-UD COMPILER, which parses the expressions of the input file and generates error messages for the user and a compressed output file to be used by PROMASA if the object definition is correct.

PROMASA (written in AutoLISP) extracts the geometric information from the output file and calculates mass properties of each component of the object with respect to the reference system defined in the input file. In addition, a wire frame model of the object is created (using AutoCAD) for visual confirmation of the object geometry. The last option in TANGO-UD main menu, the ANALYZER, is used to analyze the mass properties of the object. The user can choose a sorting criteria such as mass, volume, moment of inertia, product of inertia or radius of gyration to cause the routine to output a summary report that indicates the sensitivity of an overall body to the position and orientation of its individual components.

The user is able to choose the system units (Metric or English) adopted in calculations and the position as well as orientation of the reference system with respect to the global coordinate system. This procedure is performed by the definition commands, Metric or English, and movement commands for reference system definition.

The example in Figure 4 adopts the Metric system, where the distances are measured in millimeters (these distances are internally converted into meters for mass properties calculations). The reference system has been

Metric
ROTx 30
ROTy 40
ROTz 90
TRSx 50
TRSy -45
TRSz -32

Figure 4 Reference System and Units

rotated 30, 40, and 90 degrees about the x, y, and z axes of the global frame. Also, the origin of the reference system has been translated with respect to the global frame: 50, -45 and -32 mm along the x, y, and z axes, respectively.

When the definition of the reference system requires translation but no rotation, the command **DFLTr** (default rotation) can be used. In similar way, when the reference system is only rotated, the command **DFLTt** (default translation) may be used. When mass properties are calculated with respect to the global system (reference and global frames are coincident), the command **DFLT** (default rotation and translation) may simplify the definition.

The designer is provided with a set of four basic building blocks which are the primitive elements described earlier. Each primitive element has its own parameters: **BOX** (name, density, length, width, height), **WED** (name, density, length, width, height), **CYL** (name, density, radius, height), **SEC** (name, density, radius, height) as shown in Figure 3. Every primitive must have a mass density and a unique name.

When primitives are added together, the result is called an object. A primitive can be considered as a special case of an object. A negative object (defined by one or more primitives with negative density), describes a finite void and is only important if wholly contained within a positive density object.

A primitive can be altered by a linear transformation from the global coordinate system. Two transformations are possible: translation and rotation. Translation is expressed by a vector displacement with origin given by (**FROMx, FROMy, FROMz**) and end given by (**TOx, TOy, TOz**). Rotation is decomposed into three rotations around the coordinate axes of the global coordinate system (**ROTx, ROTy, ROTz**). Again, as defined for the reference system, there are three commands used for special cases: **DFLTr** (when the primitive is only translated), **DFLTt** (when the primitive is only rotated), and **DFLT** when the primitive is not altered by a transformation.

As an example of the command definitions given in this section, suppose that the mass properties of a box with a cylindrical hole must be calculated (Figure 5). The box has a base of 2 inches in width and 2 inches in length with a height of 4 inches. The box is displaced 3 inches along the z axis, and its longitudinal axis is parallel to y axis. The hole is 1 inch in diameter, with its centerline parallel to the longitudinal axis of the box. The object is made of steel (density is taken to be 490 pounds per cubic foot). The mass properties must be calculated with respect to a system displaced (0,20,10) inches and rotated (0,0,20) degrees respect to the global coordinate system. The definition of this object in TANGO-UD language is given in Figure 5. Reserved words (e.g. BOX) and symbols given by the user such as names (e.g. block), can be upper case, lower case or a combination of both. Line spaces between the definition of primitives are optional. The command END must appear at the termination of the

```
English

ROTx 0
ROTy 0
ROTz 20
TRSx 0
TRSy 20
TRSz 10

BOX=block
density 490
length 2
width 2
height 4
ROTx -90
ROTy 0
ROTz 0
FROMx 0
FROMy 0
FROMz 0
TOx -1
TOy 0
TOz 4

CYL=hole
density -490
radius 0.5
height 4
ROTx -90
ROTy 0
ROTz 0
FROMx 0
FROMy 0
FROMz 0
TOx 0
TOy 0
TOz 3

End
```

Figure 5 TANGO-UD Definition of a Simple Object

code. Figure 6 shows the wire frame model and corresponding coordinate systems created by PROMASA when the example above is input into TANGO-UD.

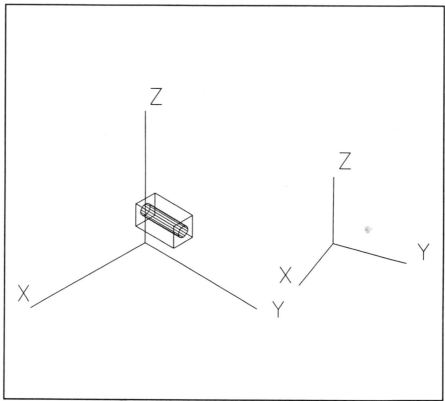

Figure 6 Wire Frame Model

VI. SAMPLE ANALYSIS

This section is intended to show in a general way the analysis of a complex welding gun unit intended for use with a commonly applied robot. The original engineering drawing is shown in Figure 7. The object was built by

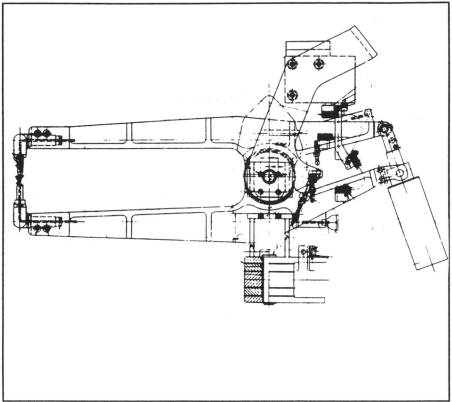

Figure 7 Welding Gun Unit

modeling the complex shapes comprising the unit with simple primitives
and using the concept of addition and subtraction of volumes. The
modelled object is composed of fifty-three primitives.

Figure 8 shows an AutoCAD wireframe drawing of the welding gun unit
automatically created by the TANGO-UD source code. The ability to
graphically view the modelled geometry is essential for checking the
accuracy of the sourcecode. Mislocated or misoriented primitives are
readily seen in the context of the isometric wireframe. The designer can
check the TANGO-UD generated wireframe for discrepancies and then
go back and edit the source code to correct the problem. The coordinate

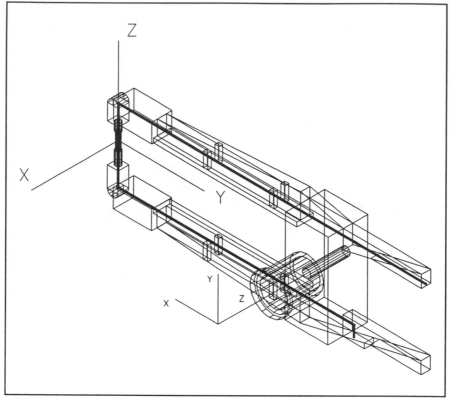

Figure 8 Welding Gun Unit Model

frames shown in the drawing represent the world frame and the reference frame.

The ultimate result of the modeling process is the moment of inertia analysis of the weld gun unit. A complete moment of inertia listing for each element modelled and the total inertia resulting from the aggregate body are produced. The designer can quickly see which elements in the body most strongly influence the overall moment of inertia.

The practical application of this information involves comparison of the output generated by TANGO-UD with the robot load specifications.

Figure 9 shows wrist load specifications for a robot used to carry this welding gun in its automobile body assembly application. The moments of inertia with respect to the reference system (coincident with wrist axis of the robot) were compared with the specifications provided in Figure 9.

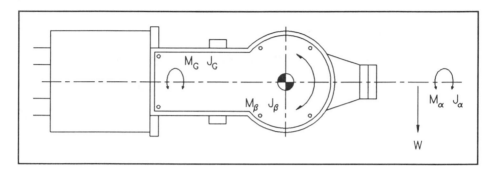

Maximum Load Capacity (W) 250 mm from Mounting Face	60 *Kg* (132 *lb*)
Moment about *G* Axis (M_G)	25.8 *Kg–m* (186*ft–lb*)
Moment about β Axis ($M_β$)	25.8 *Kg–m* (186*ft–lb*)
Moment about α Axis ($M_α$)	9.0 *Kg–m* (65 *ft–lb*)
Inertia about *G* Axis (J_G)	112.5 *Kg–cm–s²*
Inertia about β Axis ($J_β$)	112.5 *Kg–cm–s²*
Inertia about α Axis ($J_α$)	13.8 *Kg–cm–s²*

Figure 9 Robot Wrist Load Specifications

The overall results, shown in Figure 10, show that this welding gun unit can substantially overload the α axis of the robot. If the goal is to optimize the welding gun inertia distribution, a complete study of the contribution from each part to the overall inertia can be realized.

VII.   CONCLUSIONS

The main reason for undertaking this work was to implement a method for calculating the moments of inertia of robot welding guns with relative ease. Experience with the program has shown, that using primitive volumes, complex shapes can be built up in a short sequences of commands. The computational cost associated with the integral properties calculation is small, providing the high speed of response needed in very complex designs.

$$I_{xx} = 61.12 \ kg \ cm \ s^2$$
$$I_{yy} = 94.40 \ kg \ cm \ s^2$$
$$I_{zz} = 43.61 \ kg \ cm \ s^2$$

Figure 10 Overall Results of TANGO-UD Modelling of the Weld Gun

The syntax of the definition language is appropriate for practicing engineers that are accustomed to using software packages, rather than programming languages. Even more, the wire frame model of the geometry described by the language, improves the communication between the designer and the program in order to correct errors in the inertial modeling of the object. Finally, the ANALYZER, is used to understand the mass properties of the model. The user can choose a different sorting criteria to analyze the sensitivity of an overall object with respect to the position and orientation of its individual components.

REFERENCES

1. S. Marin, "General Motors Research Laboratories, Internal Report," (1986).

2. G. Hartquist, "History and Development of PADL Part and Assembly Description Language," *in* "Project Socrates Workshop Notes," Cornell University, (1988).

3. Y. Lee and A. Requicha, "Algorithms for Computing the Volume and Other Integral Properties of Solids, I, Known Methods and Open Issues," *Communications of the ACM*, 635-641, (1982).

4. Y. Lee and A. Requicha, "Algorithms for Computing the Volume and Other Integral Properties of Solids II, A Family of Algorithms Based on Representation Conversion and Cellular Approximation," *Communications of the ACM*, 642-650, (1982).

5. J. Miller, "Analysis of Quadric-Surface-Based Solid Models," *IEEE Computer Graphics and Applications*, 28-42, (1988).

6. C. Mirolo and E. Pagello, "A Solid Modeling System for Robot Action Planning," *IEEE Computer Graphics and Applications*, 55-69, (1989).

7. B. Nnaji, J. Chu, and M. Akrep, "A Schema for CAD-Based Robot Assembly Task Planning for CSG-Modeled Objects," *Journal of Manufacturing Systems*, Vol. 7, No. 2, 131-145, (1988).

8. J.R. O'Leary, "Evaluation of Mass Properties by Finite Elements," *Journal of Guidance and Control*, 188-190, (1980).

9. A. Requicha, "Representations for Rigid Solids: Theory, Methods, and Systems," *ACM Computing Surveys*, 439-464, (1980).

10. I. Braid, "Designing with Volumes," Cantab Press, Cambridge, England, 1973.

11. I. Braid, "The Synthesis of Solids Bounded by Many Faces," *Communications of the ACM*, 209-216, (1975).

12. A. Baer, C. Eastman, and M. Henrion, "Geometric Modeling: A Survey," *Computer Aided Design*, 253-272, (1979).

13.   L. Cattani, "TANGO: A Simple Solid Modelling Software for Mass Properties Analysis of Robot Welding Tools," *Master of Engineering Thesis, Mechanical Engineering Department*, University of Detroit, (1989).

14.   L. Cattani and P. Eagle, "Modelling a Robot's Tools," *Mechanical Engineering*, Vol. 112, No.6, 44-46, (1990).

15.   L. Cattani and P. Eagle, "TANGO-UD: A Simple Solid Modelling Software for Mass Properties Analysis of Robot Welding Tools," *Journal of Manufacturing Systems*, Vol.10, No.4, (1991).

16.   J. Craig, "Introduction to Robotics," Addison-Wesley, New York, 1986.

17.   I. Shames, "Engineering Mechanics," Second Edition, Prentice-Hall, Englewood Cliffs, 1967.

18.   K. Symon, "Mechanics," Third Edition, Addison-Wesley, New York, 1971.

TECHNIQUES FOR OPTIMAL OPERATION ALLOCATION METHODS IN MANUFACTURING SYSTEMS - A REVIEW OF NON-PROBABILISTIC APPROACHES

J M WILSON

Loughborough University of Technology
England

I.   INTRODUCTION

The field of optimal operation allocation methods has been a rich source for theoretical research and applications development in the manufacturing systems subject area. Starting from the early work of Manne [1] in the 1950's the subject of optimal allocation has moved forward with great rapidity in the 1960's and 1970's. Developments after the 1970's might have seemed to be entering into a rather arid phase, with more research than applications to use it. However, two major changes took place which put new momentum into research and applications and this momentum has been maintained into the 1990's and looks like continuing. The two developments were:

(a)   the classification of the complexity of algorithms for many types of problem,

(b)   the introduction of flexible manufacturing systems (FMS).

Development (a) was spurred on by the work of Garey and Johnson [2] where the terms polynomial algorithm, NP hard and NP complete were established.

Development (b) has no obvious published starting point but early reports appear in Hartley for Japanese applications [3], Hartley for other dimensions [4] and a report by Ingersoll Engineers [5].

The purpose of this chapter will be to review techniques for optimal operation allocation. The survey will be structured around the following four themes:

> modeling the problem
> types of problems:
>> scheduling
>> batching
>> routing
>> lot-sizing
>> dispatching
>> sequencing
> methods of solving the problem
> application areas.

II.   MODELING

Many structures exist for manufacturing systems, for example parallel (Figure 1), hierarchical (Figure 2).

Within these structures bottle-necks may exist. A bottle-neck, Billington [6], is a work centre which limits the rate of production of the whole manufacturing system in some way. The bottle-neck may involve one or many of the production resources such as machines or tools and provides a capacity limitation (Figure 3).

The main elements of the model of a manufacturing system will be:

> set-up time and cost
> number of machines
> processing rates of machines
> amount of work allocated to a machine
> time of operations
> cost of operations (which may come from the surrogate of time)
> number of products
> demand for products
> operations necessary to produce products
> time to move between operations and machines
> batch sizes

In addition, there will be the constraints and conditions which are pertinent to the above.

These elements arise from or have impact on a variety of aspects of manufacturing technology such as CIM, CAM, FMS, MRP2 (manufacturing resource planning), JIT (just-in-time) and OPT (optimised production technology). A good survey of the last of these three letter acronyms is contained in Rand [7].

A general review of the difficulty of manufacturing systems problems appear in Florian et al [8] which sums up the position in the 1970's.

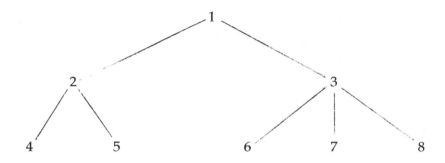

| 7 | → | 5 | → | 3 | → | 1 |
| 8 | → | 6 | → | 4 | → | 2 |

Figure 1: A parallel structure

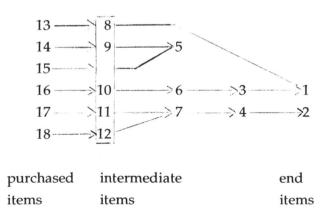

Figure 2: A hierarchical structure

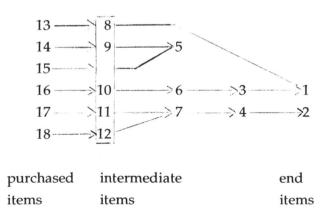

| purchased | intermediate | end |
| items | items | items |

Figure 3: A bottle-neck structure (affecting items 8-12)

III.   TYPES OF PROBLEMS

A.   Scheduling

Scheduling is defined by Bustos [9] as "the on-line definition of a timetable for the immediate future to implement a previously defined sequence". Thus scheduling concerns an overall function once particular decisions have been taken. This definition is at odds with the use of the word scheduling in job-shop scheduling problems, where scheduling essentially refers to the order in which jobs should be allocated to machines (and to which machines and possibly at which times).

B.   Batching

Batching concerns the division of the production plan into sets of parts to be made at particular shifts.

A typical FMS batching problem due to Bustos [9] could be formulated as follows.

Notation

W_{ij} index of performance for part type i on route j

X_{ij} number of parts of type i produced on route j

C_i remaining number of parts of type i to be manufactured

q_i minimum number of parts of type i to be manufactured in the next time period

t_{ijk} processing time for part type i on route j at machine group k

a_k available time of machine group k in the next time period

e_{ijk} variable which is equal to g_{ik} if machine group k is in group j and zero otherwise

y_{ij} variable which is equal to 1 if $X_{ij} \neq 0$ and zero otherwise

h_k number of free pockets in the tool-magazine k

g_{ik} number of tool slots that are necessary to manufacture
the parts of type i at machine group k in the next time
period

r_i number of possible routes for each family of parts

M   number of machine groups

N   number of types of parts

Objective

$$\text{maximise} \quad \sum_{i=1}^{n} \sum_{j=1}^{r_i} W_{ij} X_{ij}$$

Constraints

(i)   Lot-size $\displaystyle\sum_{j=1}^{r_i} X_{ij} \leq C_i, \ (i = 1, \dots, N)$

(ii)   Due date $\displaystyle\sum_{j=1}^{r_i} X_{ij} \leq q_i, \ (i = 1, \dots, N)$

(iii) Processing time $\displaystyle\sum_{i=1}^{N} \sum_{j=1}^{r_i} t_{ijk} X_{ij} \leq a_k, \ (k = 1, \dots, M)$

(iv)   Capacity $\displaystyle\sum_{i=1}^{N} e_{ijk} y_{ij} \leq h_k \ (k = 1, \dots, M)$

This model is an integer program and hard to solve when M and
N grow to significant size. Such mathematical programming
problems are typical of optimal operation allocation problems

and attempts to get round their difficulty of solution form the kernel of almost all published work.

Batch sizes are known to have a significant affect on flow times. The work of Dobson et al [10] reviews research in this area and discusses how batching policies affect various performance measures. Dobson et al [10] are able to derive various formulae which relate more in style to inventory formulae to solve batching problems. This is in contrast to the mathematical programming approaches alluded to above.

C.   Routing
Routing defines the processes through which each part-type must pass.

Using the model of Bustos [9] described in III, B, the routing problem then becomes an iterative problem. Completion time must be minimized and work balancing must be achieved until some form of optimal solution is reached. It will be desirable to have an even utilization of all machines which improves the flow of parts and to control bottlenecks.

D.   Lot-sizing
Lot-sizing is the determination of optimal sizes of production quantities once other decisions such as routing and sequencing have been taken. For unconstrained problems a large number of straightforward techniques are available involving optimal policies and heuristics such as Silver and Meal [11]. When the problem becomes constrained, e.g. by a bottleneck, more complex approaches are necessary. Usually these approaches are of the mathematical programming type and the work of Billington et al [6] is typical.

E.   Dispatching
Dispatching concern the order of the introduction of parts into
the manufacturing system.

Bustos [9] suggests that it could be argued that the dispatching
function should be included in sequencing becomes its aim is to
define the sequencing of the load operation.

However, Bustos [9] goes on to argue that two reasons suggest
that dispatching can be identified as a separate problem. The
reasons relate to:

(a)   For an FMS intervention into sequencing is usually
       precluded from control software and so the order in
       which parts are dispatched provides the opportunity to
       influence the plan of output. Thus the order of
       introduction of parts provides this control mechanism.

(b)   Dispatching only becomes a relevant operation when a
       production or manufacturing system "knows" about each
       part. (Bustos [9] terms this a boundary operation.)

F.   Sequencing
Sequencing defines the order in which operations will be
performed or may define a set of rules to be used to determine
that order. The operations typically include manufacturing,
tooling and part transporting operations.

In sequencing it is generally assumed that the allocation of parts
to machines has already been decided. Wittrock [12] reviews
work in this area and updates earlier work of Dannenbring [13]
and Lawler et al [14]. Wittrock [12] proposes a heuristic
approach to solving sequencing problems which although a

local optimization approach, has some of the style of dynamic programming.

IV. METHODS OF SOLVING THE PROBLEM

A number of authors have reviewed work on problem solving in the optimal allocation area. Goyal and Gunasekaran [15] review the connection between production and inventory. The factors involved in scheduling, batching and lot-sizing have recently been integrated in work by Potts and Van Wassenhove [16] to classify the complexity of problems and to discuss methods of solution. Their work will be discussed in detail.

They propose a general modeling approach as follows:

There are   M machines $(m = 1, .. , M)$

N jobs $(j = 1, .. , N)$, each of which is assigned to one of F families

P_{jm} is the processing time of job j on machine m

each job j contains q_j identical items each requiring processing time p_{jm}/q_j

C_i is the completion time of item i (i = item number)

c_f is the set up cost of family f

s_f is the set up time of family f

W_i is a weight on completion time

d_i = due date of job i

$U_j = 1$ if $C_i > d_i$

     $= 0$ otherwise

$T_i = \max \{C_i - d_i, 0\}$

$L_i = C_i - d_i$

Possible objectives for the problems are then:

maximum completion time　　　　$C_{MAX} = \max_j (C_j)$

total (weighted) completion time　$\Sigma_i (W_i) C_i$

maximum lateness　　　　　　　$L_{MAX} = \max_j (L_j)$

total (weighted) tardiness　　　　$\Sigma_i (W_i) T_i$

or　(weighted) number late　　　$\Sigma_i (W_i) U_i$

and all other maxima and summations are over all items i. For set-up cost models, the total set-up cost is added to the appropriate objective to give a cost function to be minimised. Potts and Van Wassenhove [16] use the three-field descriptor approach of Lawler et al [17] to classify problems, namely

machine structure/cost structure/objective

and provide the tables of classification to show batching problems with sequence-independent costs (Table I) and lot-sizing problems (Table II).

Table I. Complexity of batching problems with sequence independent set-ups

Problem	Group Technology	Set-up Times Fixed F	Arbitrary F	Set-up Costs Fixed F	Arbitrary F
$1/s_f/L_{max} + \Sigma c_f$	$O(N \log N)$	$O(F^2 N^{2F})$	NP-hard	$O(F^2 N^{2F})$	NP-hard
$1/s_f/\Sigma(w_i)C_i + \Sigma c_f$	$O(N \log N)$	$O(F^2 N^{2F})$	open	$O(F^2 N^F)$	open
$1/s_f/\Sigma U_i + \Sigma c_f$	open	$O(F^2 N^{F+1})$	NP-hard	$O(F^2 N^{F+1})$	NP-hard
$P/s_f/\Sigma C_i + \Sigma c_f$	open	open	open	open	open
$F2/s_f/C_{max} + \Sigma c_f*$	$O(N \log N)$	$O(F^2 N^{2F})$	open	$O(F^2 N^{2F})$	open
$O2/s_f/C_{max} + \Sigma c_f$	open	open	open	open	open

* Results refer to the permutation flow-shop.

Table II. Complexity of lot-sizing problems

Problem	Completion Times		
	Job	Item	Sublot
$1/q_j,t_j/L_{max} + \Sigma b_j$	$O(N \log N)$	$O(N \log N)$	$O(N \log N)$
$1/q_j,t_j/\Sigma(w_i)C_i + \Sigma b_j$	$O(N \log N)$	$O(N \log N)$	open*
$1/q_j,t_j/\Sigma U_i + \Sigma b_j$	NP-hard	NP-hard	NP-hard
$P/q_j,t_j/\Sigma C_i + \Sigma b_j$	NP-hard	NP-hard	NP-hard
$F2/q_j,t_j/C_{max} + b_j$	open	open	open
$O2/q_j,t_j/C_{max} + \Sigma b_j$	$O(N)$	$O(N)$	$O(N)$

* Solvable in $O(N \log N)$ time for set-up cost model with continuous sublots.

Problems may be solved as follows. Firstly for general scheduling problems.

A. Single machine problems

The maximum lateness problem $1//L_{MAX}$ is solved by the method of Jackson [18].

The total weighted completion time problem $1//\Sigma W_i C_i$ by the method of Smith [19].

The number late problem $1//\Sigma U_i$ by the method of Moore [20].

B. Parallel machine problems

The total completion time problem $P//\Sigma C_i$ is solved by the method of Conway et al [21].

C. Two machine problems

The maximum completion time flow-shop problem $F2//C_{MAX}$ is solved by the method of Johnson [22].

The maximum completion time open-shop problem $O2//C_{MAX}$ is solved by the method of Gonzalez and Sahni [23].

Secondly for batching problems (see Table I)

The main reference here is Monma and Potts [24] which describes dynamic programming approaches to group technology, maximum lateness and total weighted completion time problems. This expands to algorithms for two-machine flow shops in Sekiguchi [25] using the well established Johnson's rule [22].

In the three-field descriptor used in Table I, s_f in the second field denotes a sequence of independent set-up times for each family f and Σc_f in the third field indicates that set-up costs for each family f are to be considered.

Thirdly for lot-sizing problems (see Table II).
Useful references here are Santos and Magazine [26], Dobson et al [10] and Naddef and Santos [27]. As in Table I, a three-field problem descriptor is used in Table II. q_j and t_j in the second field indicate that each job contains several items and a set-up time is required for each sub-lot. For the third field Σb_j shows that a set-up cost is incurred for each sub lot.

5.   APPLICATION AREAS

This section will consider a number of recent application areas and comment on pertinent features.

A.   Tape Manufacture
A study by Markland et al [28] considers a tape (i.e. duct tape, masking tape, filament tape) manufacturing plant.   A

scheduling approach was developed jointly using heuristic approaches on micro computers and IP approaches on a large mainframe. The scheduling was required to get round plant operation problems, inventory problems and customer service problems. An IP goal programming model is developed and makes use of developments on rules from Conway et al [21].

B.   Chemical Manufacture
Selen and Heuts [29] consider a production planning problem involving a chemical reactor. Set-up time is sequence dependent and there is a single bottle-neck. An algorithm is developed to provide a heuristic approach for smaller schedules on a micro-computer. The algorithm aims to reduce holding cost and set up costs and aspects of the work related to the travelling salesman problem (TSP) (see for instance [30]).

C.   Manufacture using NC-machines
Avonts et al [31] describe an LP approach to an FMS planning problem involving six NC-machines. The LP model is fairly simple and makes partial decisions on which products and in what quantities to produce in the FMS. A micro computer based approach is adopted.

6.   SUMMARY

This review has indicated that the area of optimal operation allocation methods has been a rich one for research. As computer systems become more sophisticated but simultaneously more accessible, it is likely that implementation of research ideas will be speeded up. The development of FMS and its technological successors will both require and force the integration of research ideas and practical application.

REFERENCES

1. A.S. Manne, "On the Job-shop Scheduling Problem," Opns. Res. 8, 1960, 219-223.

2. M.R. Garey and D.S. Johnson, "Computers and Intractability," Freeman, New York, 1979.

3. J. Hartley, "Flexibile Automation in Japan," IFS/Springer Verlag, Bonn, 1984.

4. J. Hartley, "FMS at Work," IFS/North Holland, Amsterdam, 1984.

5. Ingersoll Engineers, "The FMS Report," IFS Publications, Bedford, 1982.

6. P.J. Billington, J.O. McLain and L.J. Thomas, "Heuristics for Multi-level Lot-sizing with a Bottleneck," Mgmt. Sci. 32, 989-1006, 1986.

7. G.K. Rand, "MRP, JIT and OPT," in "Operational Research Tutorial Papers (L.C. Hendry and R.W. Eglese eds.), The Operational Research Society, England.

8. M. Florian, J.K. Lenstra and A.H.G. Rinooy Kan, "Deterministic Production Planning: Algorithms and Complexity," Mgmt. Sci. 26, 12-20, 1980.

9. J.M. Bustos, "Batching and Routing: Two functions in the operational planning of flexible manufacturing systems," Euro. J. Opl. Res. 33, 1988, 230-244.

10. G. Dobson, U.S. Karmarkar and J.L. Rummel, "Batching to Minimize Flow Times on Parallel Heterogeneous Machines," Mgmt. Sci. 35, 1989, 607-613.

11. E.A. Silver and H.C. Meal, "A heuristic for selecting lot-size quantities for the case of a deterministic time-varying demand rate and discrete opportunities for replenishment," Production and Inventory Management 14, 1973, 64-74.

12. R.J. Wittrock, "An Adaptable Scheduling Algorithm for Flexible Flow Lines," Opns. Res. 36, 1988, 445-453.

13. D.G. Dannenbring, "An Evaluation of Flow Shop Sequencing Heuristics," Mgmt. Sci. 23, 1977, 1174-1182.

14.   E.L. Lawler, J.K. Lenstra and A.H.G. Rinooy Kan, "Recent Developments in Deterministic Sequencing and Scheduling: A Survey" *in* "Deterministic and Stochastic Scheduling," pp. 35-73, M.A.H. Dempster et al (eds.), D Reidel, Boston, 1982.

15.   S.K. Goyal and A. Gunasekaran, "Multi-stage production-inventory systems," Euro. J. Opl. Res. **46**,

16.   C.N. Potts and L.N. Van Wassenhove, "Integrating Scheduling with Batching and Lot-sizing," Faculty of Mathematical Studies, University of Southampton, 1990 (to appear in J. Opl. Res. Soc.).

17.   E.L. Lawler, J.K. Lenstra, A.H.G. Rinooy Kan and D. Shmoys, "Sequencing and Scheduling: Algorithms and Complexity," Econometric Institute, Erasmus University, Report 8934/A, 1989.

18.   J.R. Jackson, "Scheduling a Production Line to Minimize Maximum Tardiness," University of California, Los Angeles, Management Science Research Report **43**, 1955.

19.   W.E. Smith, "Various Optimizers for Single Stage Production," Nav. Res. Logist. Q. **3**, 59-66, 1956.

20.   J.M. Moore, "An n-job, One Machine Sequencing Algorithm for Minimizing the Number of Late Jobs," Mgmt. Sci. **15**, 102-109, 1968.

21.   R.W. Conway, W.L. Maxwell and L.W. Miller, "Theory of Scheduling," Addison-Wesley, Reading, 1967.

22.   S.M Johnson, "Optimal Two-and Three-stage Production Schedules with Set-up Times Included," Nav. Res. Logist. Q. **1**, 61-68, 1954.

23.   T. Gonzalez and S. Sahni, "Open Shop Scheduling to Minimize Finish Time," J. Assoc. Comput. Mach. **23**, 665-679, 1976.

24.   C.L. Monma and C.N. Potts, "On the Complexity of Scheduling with Batch Set-up Times," Opns. Res. **37**, 798-804, 1989.

25.   Y. Sekiguchi, "Optimal Schedule in a GT-type Flow-shop under Series-parallel constraints," J. Opns. Res. Soc. Japan **26**, 226-251, 1983.

26.   C. Santos and M. Magazine, "Batching in Single Operation Manufacturing Systems," Opns. Res. Lett. **4**, 99-103, 1985.

27.   D. Naddef and C. Santos, "One-pass Batching Algorithms for the
       One Machine Problem," Discrete Appl. Math. 21, 133-145, 1988.

28.   R.E. Markland, K.H. Darby-Dowman and E.D. Minor, "Co-
       ordinated Production Scheduling for Make-to-order
       Manufacturing," Euro. J. Opl. Res. 45, 155-176, 1990.

29.   W.J. Selen and R.M.J. Heuts, "Operational Production Planning
       in a Chemical Manufacturing Environment," Euro. J. Opl. Res.
       45, 38-46, 1990.

30.   E.L. Lawler, J.K. Lenstra, A.H.G. Rinooy Kan and D.B. Shmoys
       (eds.), "The Traveling Salesman Problem: A Guided Tour of
       Combinatorial Optimization," Wiley, New York, 1985.

31.   L.A. Avonts, L.F. Gelders and L.N. Van Wassenhove,
       "Allocating Work Between an FMS and a Conventional Jobshop:
       A Case Study," Euro. J. Opl. Res. 33, 245-256, 1988.

A BRANCH-AND-BOUND ALGORITHM
FOR SOLVING
THE MACHINE ALLOCATION PROBLEM

CHUNHUNG CHENG
ANDREW KUSIAK
WARREN J. BOE

Intelligent Systems Laboratory
Department of Industrial Engineering
The University of Iowa
Iowa City, IA 52242

I. INTRODUCTION

Group technology (GT) is concerned with the formation of part families and machine cells. The result of grouping machines and parts may lead to either a physical or logical machine layout [1]. The physical machine layout implies rearrangement of machines on the shop floor. The logical machine layout requires virtual grouping of machines and therefore does not alter the position of machines on a shop floor.

One of the frequently used representations of the group technology problem is a machine-part incidence matrix $[a_{ij}]$ which consists of "0", "1"

CONTROL AND DYNAMIC SYSTEMS, VOL. 47

entries, where an entry "1" ("0") indicates that machine i is used (not used) to process part j. Existing heuristics for solving the group technology problem in the matrix representation are as follows.

- Similarity coefficient methods ([2] and [3]).
- Sorting based algorithms ([4] and [5]).
- Bond energy algorithms ([6] and [7]).
- Cost-based methods ([8] and [9]).
- Non-hierarchical clustering algorithms ([10] and [11]).

Alternatively, mathematical programming models can be used to study the group technology problem. The models differ in objectives and constraints but are useful in rigorous and comprehensive study of the group technology problem.

Kusiak et al. [12] apply the p-median model and the quadratic programming model to group parts into a predetermined number of families. The objective function of both models is to minimize the total sum of distances between any two parts i and j. A subgradient algorithm and an eigenvector-based algorithm are proposed to solve the p-median and the quadratic programming model, respectively.

Choobineh [13] proposes a two-stage procedure for the design of a cellular system. In the first stage, parts are grouped into families using similarity coefficients based on their operations and machining sequence. In the second stage, an integer programming model is formulated for the formation of machine cells and the assignment of part families to machine cells. Intercellular moves are not allowed.

Gunasingh and Lashkari [14] propose a sequential modeling approach to the machine cell formation. First, machines are grouped into cells based on their similarity in part processing and then parts are allocated to appropriate machine cells based on the processing requirements. Although the intercellular moves are allowed in their model, they are not directly considered in the objective function for the grouping of machines.

Gunasingh and Lashkari [15] apply an integer programming model approach to the machine allocation problem. Assuming that parts have been grouped into families, several 0-1 integer programming models are formulated for the allocation of machines to machine cells. Although the models are formulated for a physical machine layout, they can be easily modified for a logical machine layout.

A 0-1 integer programming model for the machine allocation problem is formulated in this paper. Similar to Gunasingh and Lashkari [15], the new model explicitly considers the cost of allocating machines to cells and the cost of intercellular moves. Since the problem is shown to be NP-complete, the general purpose integer programming software used in previous research such as that of Choobineh [13], and Gunasingh and Lashkari ([14] and [15]) is not suitable for solving large scale problems. Special purpose algorithms which take advantage of the problem structure are therefore needed.

The problem of allocating machines to machine cells can be reduced to the generalized assignment problem by considering exactly one copy of each machine type. The generalized assignment problem is solved to provide an initial solution to the orginal problem. Assigning the remaining machines of different types to a machine cell is equivalent to solving the corresponding knapsack problem for the cell. In order to solve the overall allocation problem, an efficient branch-and-bound approach is proposed.

In the next section, the problem of allocating machines to cells is formulated and the computational complexity of the problem is discussed. The branch-and-bound algorithm for solving the problem is specified in Section 3. An illustrative example is given in Section 4. The computational results are presented in Section 5.

II. THE MACHINE ALLOCATION MODEL

The model is concerned with the allocation of machines to machine cells. Parts must be assigned to machine cells (i.e., part families) before this model can be applied. To model the problem of allocating machines to machine cells, define:

k = cell index, $k = 1, ..., p$

i = part index, $i = 1, ..., n$

j = machine index, $j = 1, ..., m$

f_j = the utilization cost of machine j

h_j = operating time required by machine j

B_j = the maximum number of copies of machine j available

d_i = the demand for part i

q_i = size of unit handling load of part i

g_i = the cost of handling a unit load of part i out of a cell to complete an operation

G_k = the maximum operating time available in cell k

$$a_{ij} = \begin{cases} 1 & \text{if part } i \text{ requires machine } j \\ 0 & \text{otherwise} \end{cases}$$

$$b_{ik} = \begin{cases} 1 & \text{if part } i \text{ is assigned to machine cell } k \\ 0 & \text{otherwise} \end{cases}$$

$$x_{jk} = \begin{cases} 1 & \text{if machine } j \text{ is assigned to machine cell } k \\ 0 & \text{otherwise} \end{cases}$$

The problem is formulated as follows:

$$\text{IP1:} \quad Z = \min \sum_{j=1}^{m} \sum_{k=1}^{p} \left[f_j - \sum_{i=1}^{n} b_{ik} (d_i / q_i) g_i a_{ij} \right] x_{jk} \qquad (1)$$

$$\text{s.t.} \quad \sum_{j=1}^{m} h_j x_{jk} \leq G_k \qquad \text{for all } k \qquad (2)$$

$$\sum_{k=1}^{p} x_{jk} \leq B_j \qquad \text{for all } j \qquad (3)$$

$$\sum_{k=1}^{p} x_{jk} \geq 1 \qquad \text{for all } j \qquad (4)$$

$$x_{jk} = 0, 1 \qquad \text{for all } j, k \qquad (5)$$

The objective function in problem IP1 seeks a tradeoff between the cost of allocating machines to machine cells and the cost of intercellular moves. When machine type j is assigned to cell k, parts in cell k requiring machine type j can be processed within the cell. Only when machine type j is not allocated to cell k, parts in cell k requiring machine type j are sent out of the cell for machining.

Constraint (2) ensures that the total operating time required by all machines in a cell does not exceed the total operating time available in the cell. Constraint (3) imposes the maximum number of copies of a machine type available for allocation. Constraint (4) ensures that at least one copy of each machine type is assigned in a manufacturing system.

The following lemma outlines the computational complexity of problem IP1.

Lemma 1. Problem IP1 is NP-complete.

Proof. Consider an instance of problem IP1 with

    1) two cells (i.e., $p = 2$),

    2) one copy for each machine type (i.e., $B_j = 1$ for all j),

    3) The term $c_{jk} = f_j - \sum_i b_{ik} (d_i / q_i) g_i a_{ij}$ is positive for all j and k.

Let $M = \max_{j, k} (c_{jk})$ and $s_{jk} = M - c_{jk}$.

The machine allocation model for cells 1 and 2 is formulated as problem IP2.

$$\text{IP2:} \quad Z = \max \sum_{j=1}^{m} \sum_{k=1}^{2} s_{jk} \, x_{jk}$$

$$\text{s.t.} \quad \sum_{j=1}^{m} h_j \, x_{jk} \le G_k \qquad \text{for all } k$$

$$\sum_{k=1}^{2} x_{jk} = 1 \qquad \text{for all } j$$

$$x_{jk} = 0, 1 \qquad \text{for all } j, k$$

To solve problem IP2, one needs to solve the machine allocation problem for one machine cell. The model for machine cell 1 is given as follows:

$$\text{IP3:} \quad Z = \max \sum_{j=1}^{m} s_{j1} \, x_{j1}$$

$$\text{st.} \quad \sum_{j=1}^{m} h_j \, x_{j1} \le G_1$$

$$x_{j1} \quad = \quad 0 \text{ or } 1 \qquad \text{for all } j$$

Then the machine allocation for cell 2 can be determined by

$$x_{j2} \quad = \quad 1 - x_{j1} \qquad \text{for all } j$$

Since problem IP3 is a knapsack problem which is NP-complete [16], one concludes that problem IP1 is also NP-complete. QED.

III. SOLVING THE MACHINE ALLOCATION MODEL

This section describes an efficient algorithm for solving the problem of allocating machines to machine cells. The algorithm first obtains an initial solution by solving a generalized assignment problem. Then the algorithm using the initial solution as a starting point employs a branch-and-bound approach based on the knapsack problem to search for the better solution.

A. OBTAINING AN INITIAL SOLUTION

Problem IP1 can be reduced to the generalized assignment problem by considering exactly one copy of each machine type (i.e., $B_j = 1$ for all j). The reduced problem is given as follows:

$$\text{IP4:} \quad Z = \min \sum_{j=1}^{m} \sum_{k=1}^{p} \left[f_j - \sum_{i=1}^{n} b_{ik} (d_i / q_i) g_i a_{ij} \right] x_{jk} \qquad (6)$$

$$\text{s.t.} \quad \sum_{j=1}^{m} h_j x_{jk} \leq G_k \qquad \text{for all } k \qquad (7)$$

$$\sum_{k=1}^{p} x_{jk} = 1 \qquad \text{for all } j \qquad (8)$$

$$x_{jk} \quad = \quad 0, 1 \qquad \text{for all } j, k \qquad (9)$$

Let $c_{jk} = f_j - \Sigma \, b_{ik} \, (d_i / q_i) \, g_i \, a_{ij}$. Since c_{jk} may be positive or negative, problem IP4 is not in the suitable form for existing solution procedures for the generalized assignment problem. Therefore the following transformation is needed.

(1) Let $M = \max_{j,k} (c_{jk})$ (10)

(2) Define $s_{jk} = M - c_{jk}$ (11)

After the transformation, the new problem can be solved as the generalized assignment problem. The transformed problem is presented next.

$$\text{IP5:} \quad Z = \min \sum_{j=1}^{m} \sum_{k=1}^{p} s_{jk} \, x_{jk} \qquad (12)$$

$$\text{s.t.} \quad \sum_{j=1}^{m} h_j \, x_{jk} \leq G_k \qquad \text{for all } k \qquad (13)$$

$$\sum_{k=1}^{p} x_{jk} = 1 \qquad \text{for all } j \qquad (14)$$

$$x_{jk} = 0, 1 \qquad \text{for all } j, k \qquad (15)$$

Several existing solution procedures such as those presented in Ross and Soland [17], Klastorin [18], and Martello and Toth [19] can be used to solve problem IP5. The heuristic algorithm developed by Martello and Toth [19] is employed in this research.

The initial solution found in this stage has two features:

1. It is a feasible solution to problem IP1 and therefore it can be treated as an incumbent solution providing a good bound for the branch-and-bound method described in the next section.

2. It helps to fix some decision variables. The branch-and-bound method begins with the machine allocation suggested by the initial solution and incorporates additional machines to cells to minimize the overall cost without violating any constraints of problem IP1.

B. THE BRANCH-AND-BOUND APPROACH

The initial solution defines a feasible solution to problem IP1. Improvements over the initial solution are possible only when

$$c_{jk} = f_j - \sum_i b_{ik} (d_i / q_i) g_i a_{ij} \leq 0$$

for some machine type j and cell k.

For machine cell k, the algorithm finds machine type j such that $c_{jk} \leq 0$. Not all machine type j with $c_{jk} \leq 0$ would be assigned to cell k because the upper bound on the operating time available in cell k cannot be exceeded and the number of copies of machine type j is limited.

Attempting to assign a machine type j with $c_{jk} \leq 0$ to cell k is equivalent to solving a knapsack problem for the cell. Many efficient solution methods such as Horowitz and Sahni [16], and Martello and Toth [20] are available. The later method is used in the implementation of the proposed algorithm.

The algorithm begins the machine allocation suggested by the initial solution and solves each cell's knapsack problem independently. For a problem of a realistic size, a solution obtained in this way is almost always infeasible (i.e., it violates constraint (4) of problem IP1). A branching scheme is used to search for the better feasible solution.

For a machine type that violates the availability constraint (i.e., constraint (4) of problem IP1), machines of the type are removed from cells one at a time. The corresponding knapsack problem for an affected cell is solved using machine types that can be assigned to the cell. In order to guarantee that at least one copy of a machine type is assigned to a machine cell, a machine type which is assigned to exactly one cell is automatically assigned to the original cell.

The developed branch-and-bound algorithm is specified as follows.

Step 1 (Initialization)

Solve problem IP4 to get an initial solution to problem IP1.

Initialize the root node by [x].

Store the initial solution as an incumbent solution [x*] and compute the upper bound Z_U.

For k = 1, ..., p construct
$$R_k = \{ j \mid f_j - \sum_i b_{ik} (d_i / q_i) g_i a_{ij} \leq 0 \}$$

For k = 1, ..., p

Define $P_k = \{j \mid x_{jk} = 1 \text{ and } \sum_{k' \neq k} x_{jk'} = 0\}$;

Define $V_k = G_k - \sum_{j \in P_k} h_j$;

Solve a single knapsack problem for cell k using V_k with machines j $\in R_k$ and j $\notin P_k$;

Set $x_{jk} = 1$ for j $\in P_k$.

Denote the solution value of [x] by Z_L.

If $Z_L \geq Z_U$, then go to Step 4 (the initial solution is final).

Initialize a new node with [x].

Step 2 (Branching)

Use the depth first search strategy to select an unfathomed node A.

Initialize [x] with node A.

Select the next machine type j such that $\sum_k x_{jk} > B_j$.

If no such machine type exists, go to step 5 (a feasible solution is found);

otherwise define $d = \{k \mid x_{jk} = 1\}$ and set $e = j$.

Step 3 (Bounding)

For $k \in d$ perform the following

Define $P_k = \{j \mid x_{jk} = 1$ and $\sum_{k' \neq k} x_{jk'} = 0\}$;

Define $V_k = G_k - \sum_{j \in P_k} h_j$;

Place e in Q_k, where Q_k is the set of machine types which are not allowed in cell k as governed by node A and its predecessor nodes;

Solve a single knapsack problem for cell k using V_k and all machine type $j \in R_k$,

$j \notin Q_k$ and $j \notin P_k$;

Set $x_{jk} = 1$ for $j \in P_k$;

Denote the new solution value of $[x]$ by Z_L;

if $Z_L < Z_U$, then

initialize a new node by $[x]$ and record the fact that x_{cj} is forced to 0; (otherwise, the current search path will not lead to a better solution than the current incumbent solution);

initialize $[x]$ with node A.

Step 4 (Stopping Rule)

If there are no more unfathomed nodes, stop; otherwise go to step 2.

Step 5 (Updating)

Denote the solution value of $[x]$ by Z_L.

If $Z_L \leq Z_U$, then

set $[x^*] = [x]$;

set $Z_U = Z_L$;

cut unfathomed nodes whose lower bounds are greater than the new upper bound.

Go to Step 2.

In step 2, machine types which violate the availability constraint are identified. Always select machine type j which maximizes $\sum\limits_{k} x_{jk} - B_j$. Ties can be broken by choosing the machine smallest number.

In step 3, new nodes are generated. The algorithm sorts all new nodes in increasing order of their lower bounds. Therefore, the depth first search strategy (in step 2) will always choose the node with the lowest lower bound in a level. This sorting provides the following improvements to the algorithm:

1. The unfathomed sibling nodes with higher lower bounds than that of the node containing an incumbent solution can be excluded from further consideration.

2. When the lower bound Z_L of a node is greater than the upper bound Z_U, the node and its sibling nodes with higher lower bound can be cut altogether.

The simplified flowchart of the branch-and-bound algorithm is shown Figure 1.

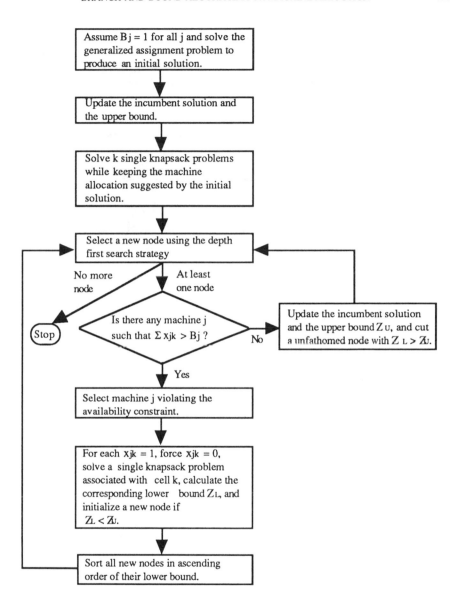

Figure 1.The simplified flowchart of the branch-and-bound algorithm

IV. AN ILLUSTRATIVE EXAMPLE

Table I.   The demands, part family assignments, and required
machining operations of parts

Part i	Demand (d_i)	Part family assignment	Required machining operations
1	32,000	3	3,4,8,9
2	96,000	2	2,3,8,9,10
3	36,000	1	1,2,5,7,9,10
4	24,000	3	1,4,5,9,10
5	26,000	1	1,2,3,5,6,8
6	18,000	2	3,4,8,10
7	34,000	2	1,3,8,10
8	32,000	1	1,2,3,4,5,6,7
9	24,000	1	2,5,6,7,10
10	22,000	3	1,3,4,9,10
11	28,000	2	1,3,4,8,9,10
12	96,000	3	1,4,8,9,10
13	46,000	1	6,7,8,9
14	78,000	3	4,8,9,10
15	22,000	2	9,10
16	72,000	2	8,9,10
17	36,000	3	8,9
18	82,000	1	7,8,10
19	68,000	1	2,5,6,9
20	96,000	3	5,6,8,9,10
21	18,000	2	1,3,5,8
22	32,000	2	2,9,10
23	20,000	1	2,5,8
24	84,000	1	2,3,4,5,8
25	72,000	1	2,3

An example is given to illustrate the proposed branch-and-bound
algorithm. The algorithm is used to solve the machine allocation problem
with 25 part types and 10 machine types. Parts have been grouped into 3
families. Table I specifies part demands, cell assignments, and operation
requirements. The size of the unit handling load for a part is 10 units. The
cost of handling a unit load of a part outside a cell for an operation is

$0.50. Table II presents the utilitization cost of machines, the number of machines available for each type, and the operating time required by each machine type. The operating time available in each cell is 1,900 minutes.

Table II.   The number of machines, the cost, and the required operating time of each machine type

Machine j	Number available (B_j)	Cost (f_j)	Operating time required (h_j)
1	2	3,000	100
2	1	8,000	400
3	2	3,000	300
4	2	4,500	600
5	1	12,000	300
6	1	11,000	200
7	1	9,000	400
8	2	2,000	200
9	1	10,500	400
10	1	11,000	100

The machine allocation problem reduces to a generalized assignment problem by restricting B_j to one for all j. The initial solution (Figure 2) which can by obtained by solving the generalized assignment problem is used to update the incumbent solution. The upper bound Z_U is -61700.

	Machine number									
	1	2	3	4	5	6	7	8	9	10
cell 1		X	X		X	X	X			
cell 2										
cell 3	X			X				X	X	X

Figure 2. The initial solution

Then a single knapsack problem for each machine cell is solved to give the results in Figure 3.

	Machine number									
---	1	2	3	4	5	6	7	8	9	10
cell 1	X	X	X		X	X	X	X		
cell 2	X		X					X	X	X
cell 3	X			X				X	X	X

Figure 3. The solution when p knapsack problems are solved

Machine types 1, 8, 9, and 10 violate their availability constraints. Branching is performed on machine type 1 first because $\sum_k x_{1k} - B_1$ is the maximum and machine type 1 is the smallest machine number. Figure 4 is a partial search tree for this problem. For the clarity of the figure, some lower bounds are shown. The order of node expansion is the order of nodes numbered. Node 1 gives the initial solution which is used as an incumbent solution and to update the upper bound. A feasible solution which is found at node 6 is stored as an incumbent solution and updates the upper bound. Nodes 10 and 14 produce feasible solutions, however, the lower bounds of the two nodes are either higher than or equal to the upper bound. The final solution is presented in Figure 5.

Solving the above machine allocation problem with the Balas' algorithm ([21]) produced the same solution. In the next section, the Balas' algorithm is compared to the branch-and-bound algorithm in solving a set of test problems.

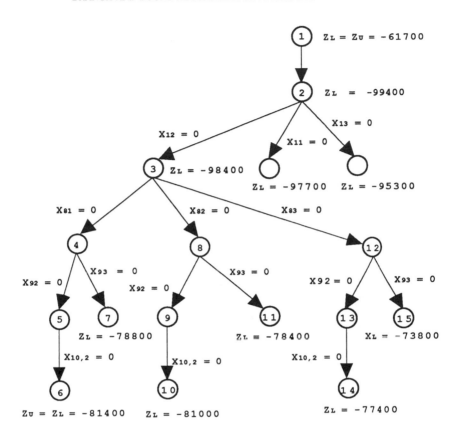

Figure 4. The partial search tree for the example

	Machine number									
	1	2	3	4	5	6	7	8	9	10
cell 1	X	X	X		X	X	X			
cell 2			X					X		
cell 3	X			X				X	X	X

Figure 5. The final solution

V. PERFORMANCE OF THE BRANCH-AND-BOUND ALGORITHM

In this section, the branch-and-bound algorithm is compared to the Balas' algorithm for a set of test problems. Both algorithms were implemented in PASCAL on a Prime 9955 computer. A code was developed to generate test problems. For each test case, twenty problems were generated. All twenty problems in a test case considered the same number of cells, machines, and parts. The number of operations, the operation requirements, and the demand for each part were randomly generated. The size of the unit handling load, and the cost of transporting a unit load of a part is fixed. The cell availiability time was fixed at half of the total time required by all machine types.

Table III summarizes the computational results. For each test case, the average CPU times for the Balas' algorithm and the developed branch-and-bound algorithm are reported. The average relative deviation in the table reports the average relative difference between the solution values of the two algorithms. The relative deviation for a test problem is defined as: 100% x (the solution value of the proposed algorithm - the solution value of the Balas' algorithm) / the absolute value of the lower solution value of the two algorithms

If the developed algorithm produces a better solution than the Balas' algorithm, the relative deviation is negative. The relative deviation is positive, otherwise. The average relative deviation for a test case is the average of relative deviations of twenty problems in the test case.

On average, the two algorithms do not generate significantly different solutions for test problems, even though the branch-and-bound algorithm seems to perform slightly better. However, the branch-and-bound algorithm is more efficient than the Balas' algorithm in most test cases.

Table III. The comparison between the developed algorithm and the
          Balas' algorithm

Number of cells	Number of machines	Number of parts	Balas' Algorithm Average CPU (sec)*	B&B Algorithm Average CPU (sec)*	Average relative deviation (%)*
3	10	20	0.05	0.06	-1.11
	15	30	0.13	0.10	-1.00
	20	40	0.29	0.13	-0.86
	25	50	0.52	0.18	-0.86
	30	60	0.56	0.21	-0.88
	35	70	1.36	0.37	-0.20
4	20	40	0.38	0.17	-0.17
	25	50	0.75	0.26	-0.35
	30	60	1.04	0.36	0.77
	35	70	1.62	0.50	-1.20
	40	80	2.73	0.69	-0.98
	45	90	4.01	0.85	0.89

*It is the average of 20 problems in each test case.

VI. CONCLUSIONS

A 0-1 integer mathematical programming model for the problem of grouping machines was formulated. In this model, tradeoffs between the cost of allocating machines and the cost of intercellular moves can be studied. The problem is NP-complete. A general purpose integer programming software is not suitable for solving realistic size problems. A branch-and-bound method based on the generalized assignment problem and the knapsack problem was developed. The performance of the branch-and-bound algorithm was reported.

References

1. A. Kusiak, "Intelligent Manufacturing Systems", Prentice Hall, Englewood Cliffs, New Jersey, 1990.

2. J. McAuley, "Machine grouping for efficient production", *Production Engineer*, 53-57, 1972.

3. H. Seifoddini and P. M. Wolfe, "Application of the similarity coefficient method in group technology", *IIE Transactions* **18**, 271-277, 1986.

4. J. R. King, "Machine-component grouping in production flow analysis: an approach using a rank order clustering algorithm", *International Journal of Production Research* **18**, 213-232, 1980.

5. H. M. Chan and D. A. Milner, "Direct clustering algorithm for group formation in cellular manufacturing", *Journal of Manufacturing Systems* **1**, 65-74, 1982.

6. W. T. McCormick, P. J. Schweitzer, and T.W. White, "Problem decomposition and data reorganization by cluster technique", *Management Science* **20**, 993-1009, 1972.

7. J. L. Slagle, C. L. Chang, and S. R. Heller, "A clustering and data recognition algorithm", *IEEE Transactions on Systems, Man and Cybernetics* **SMC-5**, 125-128, 1975.

8. R. G Askin and S. P. Subramanian, "A cost-based heuristic for group technology configuration", *International Journal of Production Research* **25**, 101-113, 1987.

9. A. Kusiak and W. S. Chow, "Efficient solving of the group technology problem", *Journal of Manufacturing Systems* **6**, 117-124, 1986.

10. M. P. Chandrasekharan and R. Rajagopalan, "An ideal seed non-hierarchical clustering algorithm for cellular manufacturing", *International Journal of Production Research* **24**, 451-464, 1986.

11. M. P. Chandrasekharan and R. Rajagopalan, "ZODIAC--an algorithm for concurrent formation of part-families and machine-cells", *International Journal of Production Research* **25**, 835-850, 1987.

12. A. Kusiak, A. Vannelli, and R. K. Kumar, "Clustering analysis: models and algorithms", *Control and Cybernetics* **15**, 139-154, 1986.

13. F. Choobineh, "A framework for the design of cellular manufacturing systems", *International Journal of Production Research* **26**, 1161-1172, 1988.

14. R. K. Gunasingh and R. S. Lashkari, "The cell formation problem in cellular manufacturing systems -- a sequential modelling approach", *Computers and Industrial Engineering* **16**, 469-476, 1989.

15. R. K. Gunasingh and R. S. Lashkari, "Machine grouping problem in cellular manufacturing systems -- an integer programming approach", *International Journal of Production Research* **27**, 1465-1473, 1989.

16. E. Horowitz and S. Sanhi, "Fundamentals of Computer Algorithms", Computer Science Press, Maryland, 1978.

17. G. T. Ross and R. M. Soland, "A branch and bound algorithm for the generalized assignment problem", *Mathematical Programming* **8**, 91-103, 1975.

18. T. D. Klastorin, "An effective subgradient algorithm for the generalized assignment problem", *Computers and Operations Research* **6**, 155-164, 1979.

19. S. Martello and P. Toth, "An algorithm for the generalized assignment problem" in "Operation Research 1981" (J.P. Brans, ed.), North-Holland, Amsterdam, 1981.

20. S. Martello and P. Toth, "An upper bound for the zero-one knapsack problem and a branch and bound algorithm", *European Journal of Operational Research* **1**, 169-175, 1977.

21. M. M. Syslo, N. Deo, and J. S. Kowalik, "Discrete Optimization Algorithms with Pascal Programs", Prentice-Hall, New Jersey, 1983.

INDEX